Programme Procurement in Construction

Programme Procurement
in Construction
Learning from London 2012

John M. Mead
Part of an International Consultancy, engaged as
Principal Programme Supply Chain Manager for the delivery of Crossrail

Stephen Gruneberg
Reader, School of Architecture and the Built Environment
The University of Westminster

WILEY-BLACKWELL
A John Wiley & Sons, Ltd., Publication

Library of Congress Cataloging-in-Publication Data

Mead, John M. (Programme manager)
  Programme procurement in construction : learning from London 2012 / John M. Mead, Part of an International Consultancy and at the time of writing engaged as Principal Programme Supply Chain Manager for Crossrail & Stephen Gruneberg, Reader, School of Architecture and the Built Environment, The University of Westminster.
       pages cm
  Includes bibliographical references and index.
  ISBN 978-0-470-67473-4 (pbk. : alk. paper) – ISBN 978-1-118-59744-6 (online product) (print) – ISBN 978-1-118-59745-3 (emobi) (print) – ISBN 978-1-118-59746-0 (epub) (print) – ISBN 978-1-118-59747-7 (epdf) (print)   1. Contractors' operations.   2. Construction industry–Management.   3. Construction contracts–United Kingdom.   4. Public works–England–London.   5. Olympic Games (30th : 2012 : London, England)   I. Gruneberg, Stephen L.   II. Title.
  TA210.M43   2013
  343.4207′8624–dc23
                              2012047263

Cover image courtesy of John M. Mead
Cover design by His and Hers Design

Set in 10.5/14 pt Trump Mediaeval by Toppan Best-set Premedia Limited
Printed and bound in Malaysia by Vivar Printing Sdn Bhd

1   2013

Contents

About the authors

John M. Mead is part of an International Consultancy and has previously led the Supply Chain Management function for the delivery of the infrastructure for the London 2012 Olympic and Paralympic Games. At the time of writing this book, he was engaged in a similar role on the £14bn Crossrail construction programme to build a railway which tunnels under London.

Stephen Gruneberg is a Reader at the University of Westminster and a Visiting Fellow at the Faculty of the Built Environment, Northumbria University. Stephen is an industrial economist specialising in the construction and property sectors and has written and co-written numerous books, reports and papers.

List of figures and tables

Forewords

Rarely, if ever, has the British construction industry delivered a major construction programme on the scale of the London 2012 Olympics, with so much success in terms of timeliness and quality while remaining within budget and without any of the acrimony that is so often associated with major construction projects. The London 2012 construction programme and its outcome have already proved to be a source of pride and achievement not only for the construction client, the workers who built it, the firms involved and the UK construction industry in general but also for the UK as a whole. So much so that politicians from all parties have praised the delivery and construction of the games infrastructure and have been keen to promote the methods used to build the Olympic Park and the other venues in order to repeat the success of the construction process and promote the methods throughout the construction industry.

Naturally, many factors came together to build that success but one of the key foundation blocks to the delivery of London 2012 was the way in which the buildings, stadia and infrastructure of the Olympic Park were procured. In my report, *London 2012 – a global showcase for UK plc*, I make the recommendation that, 'Government should adopt the principles of the procurement and programme management approach used by the ODA, [the Olympic Delivery Authority], for all public sector projects valued at over £10m'.

This book gives an account of the actual methods that were used in the programme procurement. The techniques described in this book were combined in such a way that those who led the procurement team named their approach Purchase and Supplier Engineering (PSE), an approach that could only have been developed because the complexity of the programme was recognised by the ODA, who recognised the need to allow a highly gifted and professional team to devise the most appropriate procurement methods.

This book describes the procedures, techniques and methods that were used to such great effect. It can only be hoped that the success of the London 2012 programme will be used to encourage others to adopt or adapt the same or very similar strategies, not only in the UK but around the world, wherever major construction programmes require the management and co-ordination skills and understanding that procured and delivered the London Olympics. Indeed this has already begun to happen as PSE has been adopted on the Crossrail programme, also discussed in this book. The London 2012 programme was a success; I hope this book will help future programmes to enjoy similar success.

Sir John Armitt CBE, Chairman, Olympic Delivery Authority

★ ★ ★

The delivery of the London Olympic Park and venues has won wide acclaim as a showcase for UK plc. However, at the start of this epic journey in 2006 the horizon was very different, as demonstrated by an almost total lack of interest from industry in participating.

This is not surprising as the problems of Wembley were unfolding at the time and industry enjoyed a boom of activity. Why would anyone wish to participate in such a high profile enterprise with a fixed and very public end date and an international history of delays, cost overruns and lost reputations?

Into this arena stepped the Olympic Delivery Authority as the Government agency charged with delivery. If national objectives were to be met then industry resources would have to be mobilised on a massive scale. To achieve this, the ODA would have to gain in months what many client organisations pursue for years – their establishment as a recognised client of choice.

The first stage in this process was to carefully listen to industry and its aspirations and to convert these into well defined and communicated commitments as to how the ODA, as client, would behave and how industry was expected to reciprocate. Of course, talking and doing can be very different. However, early engagement in the procurement process demonstrated that commitments would be met and built with a solid foundation on which to go forward.

Industries do not create markets; that is for clients to do and they usually get what they deserve. In the case of London 2012 the ODA received the total support and commitment of UK plc which over time developed into a matter of national pride with all participants giving of their best and no one prepared to let the side down.

It is said that those who ignore history are destined to relive it. However, the success of London 2012 does not have to be a one time achievement but rather a beacon of how things can be. This book charts how all this was achieved and for those who take note of the many lessons learned and apply them appropriately the potential for a successful outcome is vastly increased.

Howard Shiplee CBE, Executive Director, Laing O'Rourke and previously Director of Construction, Olympic Delivery Authority

⋆　⋆　⋆

The construction industry has always been a contradiction whereby those that have the vision for how our built environment should develop are trained artistically (as creative people) and those who create that environment are trained as technicians (adapters). Creative people have vision, whereas adapters solve practical problems. There is a gap. The technocrats do not fill the gap; they just help define its edges.

This book describes a methodology that fills the gap, that takes the vision of the creatives and provides a sound and tested base on which the adapters can work. Purchase and Supplier Engineering (PSE) was brought together from best practice applications across industry, the catalyst being the London 2012 Olympic Games. The need for Usain Bolt to tie up his running shoe laces at a given moment, on a specific day five years in the future, galvanised thought since 2006. The degree of national embarrassment as a result of a late Olympics was unimaginable. The result was the creation of a hugely successful methodology of procurement and supplier management that was made up of parts that in themselves were nothing particularly new, rather a great recipe made up of sound ingredients.

Crossrail is one of the largest and most significant transport projects ever undertaken in the Western hemisphere. It has little of the kudos of an Olympic Games but at an engineering level it is astonishing. To bore twin tunnels to take a full size railway under the quite ancient city of London and beyond is the stuff of HG Wells and science fiction. Crossrail has been a programme that has not played the 'not invented here' card but has instead embraced best practice. The PSE approach to procurement has been fully adopted by Crossrail Limited and has been used to procure the entirety of the portfolio of construction and engineering contracts, roughly twice the size of the Olympic Delivery Authority's programme, albeit not nearly as diverse or unique.

At the time of writing this, I cannot imagine that Crossrail will have quite the Topping Out ceremony that The London 2012 Olympics build enjoyed. However, what it does prove is that the construction legacy of the Olympics is alive and well and thriving.

Martin Rowark, Procurement Director, Crossrail

* * *

PODIUM, the Further and Higher Education Unit for the 2012 Games, was established in 2007 to maximise the engagement of universities and colleges across the United Kingdom in London 2012 and to use the Games to promote and showcase the contributions made by UK universities and colleges nationally and globally. These contributions to the legacy of London 2012 include the timely critical appraisal of all the dimensions of the Olympipc and Paralympic Games, drawing lessons for the development and management of future 'mega' events. *Programme Procurement in Construction: Learning from London 2012* by my colleague Stephen Gruneberg and his co-author John Mead is one of the first tangible academic contributions to the legacy of the Games.

The scale and complexity of the construction programme was immense; its delivery was inspiring and a global showcase for UK plc. The leader of the ODA, Sir John Armitt, has suggested that lessons should be translated to other major public sector projects. The procurement processes were at the heart of the success story. This book provides an appraisal of those processes and highlights the lessons to be learned. It will become an important element of the London 2012 legacy.

Professor Geoffrey E. Petts, Chair, PODIUM, and Vice Chancellor,
The University of Westminster

Preface

The construction of the venues and necessary infrastructure to stage the London 2012 Olympics was such a resounding success that it boosted not only the reputation of the UK construction industry, but also the confidence of the UK population in the country's ability to organise, build and run a major international event. Just as the US man-on-the-moon rocket programme challenged the competence of the whole industrial and technological base of the United States, the 2012 Olympic programme demonstrated the ability of the UK construction industry to provide a built environment to the highest standards of quality, on time, without a single fatal construction accident, in spite of its immense scale and engineering and logistical complexity.

Many factors contributed to this achievement and one of those, in particular, was the method used to mobilise the construction industry to respond to the requirements of the construction programme in the first place. Many issues needed to be resolved. How does one buy the stage for an Olympic Games? How does one engage with the construction industry that will be charged with the delivery? How does one manage the details of thousands of contracts and the many firms of contractors, subcontractors and material suppliers and ensure that no one organisation adversely affects any other, to the detriment of the programme? How does one judge quality at the tender stage? Or monitor progress? Or, for that matter, how does one measure programme exposure, or manage performance? How does one maximise competition during procurement without stepping on a legal minefield of obligations? How does one measure capacity and the ability of firms to cope with the work and the risks involved? These and many more questions and issues are dealt with in this book.

The careful and painstaking preparation of the procurement processes is discussed, ranging from understanding and developing the appetite of contractors and encouraging them to engage with the procurement process, to monitoring performance based on the contractors' own performance claims as set out in their tender submissions. The emphasis of the approach described is based on a close attention to detail to avoid surprises, while keeping a focus on the total programme. By not doing anything radically innovative or indeed difficult, but by doing simple things thoroughly in a coordinated and strategic way, a big picture is produced that is relatively easy to manage and control, with fewer Rumsfeldian unknowns.

Taken together, the methods and processes described here define an approach the authors and originators have termed Purchase and Supplier Engineering (PSE). Although similar to Supply Chain Management in many ways, PSE is a particular strategic approach to procurement that takes into account the state of the construction market at the inception of the procurement process and the early engagement of possible contractors. PSE provides an overview of the interest of firms in participating and the resulting capacity and workloads of all suppliers, including the main contractors and the critical subcontractors and materials suppliers. Having established the strategic approach for programme procurement and having organised the tendering process and awarded the contracts, PSE follows through by monitoring progress and risk throughout the construction phase and for all critical suppliers in the supply chain.

In a recent article, Hunter (2012) refers to new legislation in the UK: the Public Services (Social Value) Act, which from 2013 requires public-service contracting authorities to take into account economic, social and environmental impacts, balancing price, quality and social value in their procurement strategies. He also talks about engaging with the market before commissioning work, but he does not say how these issues may actually be dealt with in practical terms. This book describes how these and many more objectives were addressed in two major programmes: namely, the London 2012 Olympics construction programme and the Crossrail programme. The former bore the title of the largest construction programme in Europe, only to be beaten by the latter, with a combined total budget well in excess of £20bn.

The concepts and techniques used in the programme were not one-off techniques developed for the London Olympics alone, delivering success for just the ODA. Since the completion of London 2012 they have been further developed by its originators and either wholly or partially adopted on numerous other major construction programmes including the £14.8bn Crossrail rail programme, which involves tunnelling 21km of twin-bore tunnels under the heart of London and includes eight new sub-surface railway stations.

It takes a client with great vision and foresight to commit to investing in this PSE strategic approach to procurement and supply chain management, but the approach has demonstrated its ability to deliver a high degree of predictability and give clients what they set out to achieve not only in terms of the financial and economic objectives of budgets and schedules, but also in terms of the client's social and environmental objectives. The approach has also shown its ability to avoid the costs associated with supply chain insolvency, while achieving savings in common components and commodities.

We realise that these are extraordinary claims and that many factors contributed to the success of London 2012. However, in the euphoria of the Games and in the aftermath of the events, there has been very little criticism – if any – of the way the construction programme was conducted. Indeed, since the use of PSE by the ODA, elements are now being used to a greater or lesser extent on all manner of construction programmes, including those in the energy-generation, transport and utilities sectors across both public- and private-sector procurement.

For these reasons this book is aimed at public- and private-sector clients, developers, senior management and those professionals involved in undertaking the procurement, supply chain management and delivery of complex major construction programmes or those organisations, such as major tier 1 contractors, that manage large and diverse portfolios of projects across multiple client bases. The concepts described can be applied in part or in whole to portfolios of projects on a smaller or larger scale than that of an Olympic or Crossrail programme. However, to demonstrate the usefulness of the PSE model, these two particular programmes are used throughout as examples.

The specific management processes that were used in developing the PSE approach to procurement for the Olympic programme for London 2012 are described in this book. No attempt has been made to make a critical evaluation of the processes. Possibly, over time, a critique of the methods used to procure the built infrastructure of the Games may emerge. In the meantime, the authors have endeavoured to give an account of each element in the procurement process to provide the rationale behind the methods used.

Success and failure are often the result of a number of different factors. Attributing success to one particular aspect of a large and complex programme such as the Olympics can be misleading. All those involved in the many aspects of the programme were very aware of the great debt owed to the many thousands of people who contributed their diverse skills and expertise to the success of the overall project. They were also aware that numerous voices were raised in criticism of the Games, and doubts were raised about the ability of the ODA to complete their task. However, once the actual sporting events commenced, it became clear that the UK construction industry had delivered on its promises and given the London Organising Committee for the Olympic Games (LOCOG) its internationally recognised venues and infrastructure. How these promises were achieved is described in the chapters of this book, in the hope that the lessons learned from this experience can be transferred to other construction programmes and portfolios of projects both in the UK and around the world.

The book is divided into three main sections. The first part is concerned with engaging the supply market. The second part deals with the organisational aspects of programme procurement, including the appointment of contractors and the approach used to decide on the contractual arrangements used. The third part describes the management and monitoring of the performance of the critical supply chain organisations during the construction phase. The first chapter introduces the concept of Purchase and Supplier Engineering, (PSE), which was developed in response to the complexity of the programme's many procurements as their associated processes emerged. Chapter 1 discusses the structure of the organisation of the programme procurement, showing how the delivery partner was engaged as a specialist procurer on behalf of the ODA. One of the key themes of this chapter is the role of the delivery team in interpreting the goals of the client in terms of what the construction contractors were required to deliver. Delivering the vision of the client was central to the purpose of the delivery partner. The chapter concludes by introducing the distinction between projects and programmes.

Chapter 2 provides an overview of the basic economics theory that underpins PSE. That involves a clear perception of the construction market and the wider market forces facing the purchasers and suppliers. The engagement of firms is directly related to the theoretical perspective described. This approach prepares the strategy adopted in PSE for delivering procurement on large construction programmes, a strategy that creates a competitive environment amongst suppliers at all levels in the programme with the purpose of delivering value.

Chapter 3 describes the aims of the client in terms of their values, priorities and critical success factors. In particular, the needs of a client may extend beyond the provision of a physical structure, as major building programmes such as London 2012 tend to impact on local and even national economies, with wider social, environmental and political implications. The client's priorities can then be translated into the requirements to be met in the form of built structures, how they are delivered, how they are to be utilised and the legacy they leave behind over the life of the assets. The values expressed in the priorities of the client, including political priorities and often contradictory aims and ambitions, are linked to strategic goals, which in turn form the basis for measuring delivery performance and the programme's wider success.

Chapter 4 deals with the need to reduce the complexity of the many different projects within a single programme. As with any large construction project, packaging strategies have to be devised that define the parameters of each contract. In a programme on the scale of London 2012 the packaging strategy included definitions of the different facilities and the infrastructure that were needed.

Complexity is dealt with by clustering the major packages in a programme, where common characteristics can be identified. The six clusters of the London 2012 programme were structured from the ground up, much like the layers of a cake with the landscape and public-realm cluster as the icing on top. Clustering projects limited the number of different contracting solutions that were required to be used, simplifying the management of the whole process. On the London 2012 programme, most of the contract arrangements were based on variants of the third edition of the New Engineering Contract, and the chapter explains the specific contract types that were used and their appropriateness for each element in the programme.

Chapter 5 discusses the benefits of procuring common components and commodities, where these were being used across the programme or on several projects or facilities being constructed simultaneously or even concurrently. The question of whether a common component strategy is required is addressed and the chapter explains how it may be achieved, if appropriate. The advantages of purchasing common components are numerous and include securing supply, achieving economies of scale through bulk purchasing, and maintenance cost reductions.

It is essential that potential suppliers have confidence in the professionalism, knowledge and experience of the client body. With a procurement strategy in place for the facilities and the common components, the client is then in a position to approach suppliers. Chapter 6 deals with the methods used to engage with the supply market. The appetite of firms to become involved and motivated determines the capacity of the market to supply what is needed to deliver the requirements of the programme. Dialogue with suppliers is essential and no fewer than 10 different methods used are described that together generated the intelligence to inform procurement, avoid surprises and gain the commitment needed on all sides to deliver the full programme meeting the client's needs and priorities.

In Chapter 7 the PSE concept is used to standardise the procurement process and make it systematic, efficient and more of a procurement production machine. This represents the formal engagement of contractors and suppliers based on a set of standardised documents and procedures. It is essential that the tendering process is conducted efficiently with clarity, auditability, transparency and fairness. On the London 2012 programme electronic tools for eSourcing and eEvaluation were used to facilitate the tendering process via the internet. The use of digital technology fits the conventional procurement process, while standardising the approach to facilitate ease of procedural management, assurance and governance.

The relationships between the client body and contractors need to be managed throughout the tendering process and subsequently on an ongoing basis, and this aspect is discussed in Chapter 8. It is essential if the client is to maintain control over the construction process without impacting on the formal legal roles, rights and duties of the various parties engaged in the process. The client, for example, needs to map the supply chain to monitor the vast number of contracts and contractors involved. The method used to map the elements of the supply chain is shown, from directly contracted tier 1 suppliers to the indirectly sub-contracted critical tier 2 and 3 supply chain.

It is not possible for a client to procure a large programme without a system of checks to assess the reliability of the supply chain. This is known as 'due diligence' and is concerned with assessing the financial capability and capacity of firms to deliver what is required of them at project and programme levels. That leads into a discussion of risk and exposure. Chapter 9 is therefore concerned with the methods used to measure the ability of firms to deliver. Due diligence on the part of the client extends to monitoring the financial strengths of contractors relative to their contractual commitments, with a view to anticipating difficulties and dealing with them in good time, where possible and appropriate. This overview is not concerned only with main or tier 1 contractors, but is also needed at the critical tier 2 and 3 subcontractor level – and indeed applies to any supplier deemed to be of strategic importance to the programme.

In the final chapter, Chapter 10, the performance of firms in the construction phase is reviewed in terms of the commitments made by the winning suppliers during the tender process. After all, their appointment was made on the basis of the approaches described in their tender submissions. The use of the Balanced Scorecard to establish the client's priorities and the response of suppliers at tender stage to meet those requirements can therefore be compared to the actual outcomes, behaviour and processes of these firms during the delivery phase. The use of benchmarking and key performance indicators as tools to monitor progress and improve performance is also discussed.

In Chapter 1 a model of PSE is shown in Figure 1.7. We use the same diagram below, superimposing the chapter boundaries on the PSE model, to help the reader to identify the specific elements of each chapter and to make the navigation of the book relate to the logic of the deployment of PSE. In Preface Figure 1, each chapter deals with one key part of the PSE model, as described in the chapter summaries above.

Many of the key proponents and originators of the concepts deployed by the Olympic Delivery Authority (ODA) that have since come to be known and recognised as PSE contributed to this book. While there is not enough space to mention the names of the many thousands of people and organisations that came together to make London 2012 the huge success it was, the authors would like to give special mention and thanks to those key originators who developed and implemented PSE

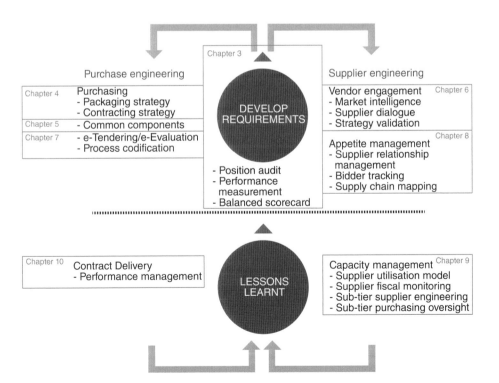

Preface Figure 1   The PSE model in relation to the chapters of this book.

on London 2012; they were Martin Rowark, Kevin Lloyd-Davies, Peter Sell and Andrew Garbutt. The authors would like to express their gratitude for their expert input and advice in developing the thinking behind the concepts described and delivered for the London 2012 programme and in bringing this book to fruition.

The authors are grateful to the Olympic Delivery Authority (ODA) and their Delivery Partner, CLM, for being forward-thinking enough to embrace the approach developed by the team for the London 2012 programme; and again we mention specifically Mark Reynolds, Kenneth Owen and Paul Dickinson of CLM and Morag Smith (née Stuart), Howard Shiplee, Huw Edwards and Sir John Armitt of the ODA for their support, both during construction and since, in recognising and championing the approaches developed.

We are also grateful to Crossrail for their support and encouragement in furthering the concepts developed for London 2012, and in particular thanks should also go to Martin Buck, Valerie Todd, and Andrew Wolstenholme. In mentioning the ODA, CLM and Crossrail, we apologise for not mentioning everyone by name directly.

Thanks should also be given to Rob Garvey for bringing the authors together and sparking the idea for this book. In addition, the authors wish to thank Madeleine Metcalfe and Beth Edgar for the time, effort and care they put into mentoring us throughout the process of writing and producing the book and to Ruth Swan, Teresa Netzler and Patrick Roberts for their vital contributions. It is, of course, in spite of their best endeavours, that responsibility for any errors and omissions remains with the authors. Recognition is also due to the the PSE team members who have helped road test and evolve this model: these include Lee Taylor, Mark Lythaby, Vanessa Good, Jitendra Chouhan, Gary Wright, Nazir Fard, Simon Pain and Joanna Lewis, all good colleagues and friends.

Finally, but by no means least, thanks are due to the authors' families for their support and patience, when spare time that was meant to be family time became more time spent working!

John M. Mead
Stephen Gruneberg

Reference

Hunter, D., (2012) 'Real value demands a bold approach', *Public Servant*, September, p. 24. http://www.publicservice.co.uk/feature_story.asp?id=20712 (accessed 5 October 2012).

List of acronyms

BAA	British Airports Authority
BMX	Bicycle motorcross
CCJ	County Court Judgement
CD-ROM	Compact disc-read only memory
CE	Constructing Excellence
CECA	Civil Engineering Contractors' Association
CH2M Hill	Cornell, Howland, Hayes, and Merryfield (CH2M Hill is the name of a firm)
CIOB	Chartered Institute of Builders
CLM	CH2M, Laing O'Rourke, Mace, the delivery partner
CPA	Construction Products Association
CSF	Critical success factors
DMIAC	Define, measure, analyse, improve, control
DNA	Deoxyribonucleic acid
DP	Delivery partner
DQI	Design quality indicator
DVD	Digital versatile disc
ECI	Early contractor involvement
EOI	Expression of interest
EPC	Engineering, procurement and construction
EU	European Union
ICT	Information and communications technology
IT	Information technology
ITT	Invitation to tender
KPI	Key performance indicator
LAD	Liquidated and ascertained damages
LEIA	Lifts and Elevators Industry Association
LOCOG	London Organising Committee for the Olympic Games
M and E	Mechanical and electrical (construction)
MEAT	Most economic advantageous tender
MS	Market sounding
NAO	National Audit Office
NEC	New Engineering Contract
NSCC	National Specialist Contractors' Council

OCI	Optimised contractor involvement
ODA	Olympic Delivery Authority
OGC	Office of Government Commerce
OJEU	Official Journal of the European Union
ONS	Office for National Statistics
PAQ	Pre-assessment questionnaire
PCR	Public contracts regulations
PFI	Private Finance initiative
PMO	Programme management office
PQQ	Pre-qualification questionnaire
PSE	Purchase and Supplier Engineering
RACI	Responsible, accountable, consulted, informed
RIA	Rail industry Association
RIBA	Royal Institute of Architects
RICS	Royal Institution of Chartered Surveyors
SCM	Supply chain management
SEC Group	Specialist Engineering Contractors' Group
SME	Small and medium sized enterprises
TES	Tender event schedule
TfL	Transport for London
WBS	Work breakdown structure

1

Purchase and Supplier Engineering and the London 2012 Olympics

Some of the millions of people who visited the Olympic Park during the Games give a sense of scale to its structures (photo courtesy of AECOM).

Programme Procurement in Construction: Learning from London 2012, First Edition.
John M. Mead and Stephen Gruneberg.
© 2013 John Wiley & Sons, Ltd. Published 2013 by John Wiley & Sons, Ltd.

Some of the 56 km of timber being installed to form the Velodrome's track (photo courtesy of Mark Lythaby).

One of the many art installations to be found in the Olympic park during the Games – this is one of the dissected and reassembled telephone boxes (photo courtesy of AECOM).

Introduction

The Games of the 30th Olympiad, held in London during the summer of 2012, gave the UK international exposure. The construction of the Olympic Park in east London and all the other Olympic venues around the country continues to receive critical acclaim. The construction of the Olympic infrastructure just had to be delivered on time; no Games had ever been delivered late and London was certainly not going to be the first. After six years of work the delivery of the 'greatest show on earth' was heralded as a triumph. This book describes how certain elements of the construction programme required to stage the Games contributed to that triumph.

In this opening chapter we therefore introduce the concept of Purchase and Supplier Engineering (PSE) and highlight the key distinguishing features of PSE from other, more standard approaches to programme procurement and supply chain management.

The Olympic Games requires the host country to deliver a spectacular stage on which the Games can be played. The scope and specifications for the infrastructure requirements of the London 2012 Olympic and Paralympic Games were no different from the 29 other Games that had preceded them. They were both diverse in engineering terms and highly complex and they were constrained by an immovable deadline for delivery – namely the opening ceremony. The project had to be completed on time and to budget, while also delivering against the legacy and environmental commitments described in London's bid proposal to host the 2012 Games. What made this challenge especially difficult was not only its scale, but that it involved almost all construction disciplines and required them to respond and overcome numerous design, engineering and construction problems. Moreover, a large and visible part of the delivery concerned the construction of sporting stadia at a time when the construction of Wembley Stadium had encountered major time and cost overruns, to the detriment of all concerned, which perhaps added to the initial reluctance of some firms to engage on the Olympics programme.

The scale of the Olympic Delivery Authority (ODA) work to deliver the stage for the London 2012 Games was valued at around £9.2 billion. Delivery of this massive investment required the procurement of approximately 2000 separate contracts in a period of less than five years. The number of contracts was comprised of approximately 250 major, large-scale construction contracts (tier 1 contracts), with the bulk of the remainder consisting of predominantly smaller procurements.

The approach adopted to procure the infrastructure required to stage the London 2012 Olympic Games has received praise from various quarters. For example, it was commended by the ODA Chairman, Sir John Armitt, as a model for other programmes to adopt. His report, 'London 2012 – a global showcase for UK plc' (2012), published on completion of the construction programme, highlighted the approaches to supply chain engagement and procurement as key contributing factors to the success of the Olympic delivery programme. The Constructing Excellence report 'Never Waste a Good Crisis', published in 2009 following the global 'credit crunch', stated that: . . . *the ODA procurement model needed to be captured and promoted not just in the UK, but around the world.*

The construction of the London 2012 infrastructure stands out as one of the largest and most successful construction projects ever undertaken in the UK. The delivery model and programme management techniques used were both innovative and robust and helped to place the UK construction industry at the forefront of international construction achievements.

The procurement model that was developed for the Olympic delivery has since been consolidated into a model for procurement and supply chain management called Purchase and Supplier Engineering (PSE). The contents of this book are structured to follow the PSE model in its chronological order, or at least as close to that as is possible. A programme by its very nature is so vast that, while the starting point for PSE is the same as for any project, many aspects that follow are likely to occur simultaneously on the many project and package procurements and contracts that make up the programme.

The concept of Purchase and Supplier Engineering

Purchase Engineering and Supplier Engineering are the two key constituent parts that make up the PSE model. The term 'Purchase and Supplier Engineering', or PSE, was introduced after the concept was developed and deployed on the London 2012 programme. In later chapters the constituent parts that make up the complete concept of PSE are described in more detail. By way of simple introduction PSE can be described as being similar to procurement and supply chain management. However, in order to highlight PSE's specific features, compared to the more traditionally held procurement and supply chain management (SCM) approaches, the application of the PSE model is described to illustrate the overall concept that has now been deployed on two of the largest construction programmes delivered in the UK in modern times – namely, the London 2012 infrastructure programme at over £9 bn and the Crossrail railway programme at some £14 bn.

PSE is a tried and tested approach, used effectively on a number of construction programmes to deliver a successful built-environment solution from a global supply marketplace. It is an approach that delivers the values and goals of the client across multiple projects within a highly complex programme or portfolio environment. The PSE approach sets out at the earliest stages to deliver the values and goals of the client.

While the PSE model was developed in response to the challenges of delivering the London 2012 Games' infrastructure, since its deployment on this programme it has been further developed and many elements of the model deployed on numerous other programmes of varying sizes and complexity.

Major construction projects and programmes are by their nature extremely multi-faceted, involving the mobilisation of resources employing complicated technical solutions. Nevertheless, this complexity can be managed by breaking the whole, regardless of its size, into appropriate and manageable parts and solving each of them individually. No overarching approach can be applied to meet all the requirements of a complex client and its many stakeholders' contradictory objectives and conflicting interests. However, it is possible to propose a consistent method for analysing each package of work and the capability and capacity of the many firms required to deliver these to meet the priorities set out for their particular part of the project.

Doing so ensures that each firm is capable of fulfilling its obligations to the project and, taken as a whole, all the firms then contribute towards the successful completion. This applies equally to the whole construction value chain, made up of main contractors, their subcontractors and their suppliers. PSE is a technique that can be used to measure and establish the capacity, capability and reliability of each firm in the supply chain even prior to selection. As procurement is not only about appointing contractors but is also very much a starting point in the process of delivery, PSE also deals with managing and monitoring certain aspects of the firms throughout their engagement.

Programme management is not the same as *project* management. Programme management is therefore definitely not the sum of project management multiplied across a number of projects in a programme. Therefore, because procurement on a programme cannot be about procuring each project in isolation, a much more strategic approach is required. The impact of one project on the total programme must be considered in the context of the whole. In a programme made up of, say, ten projects, nine successful projects would not equate to a successful programme. Therefore, PSE is not just concerned with each project; it adopts a far more strategic view of the programme's procurement and the capability of the supply chain to meet the programme's requirements.

PSE is based on the understanding that procurement is about buying a supply chain, and not just an individual contract. PSE also takes into account the wider supply issues, such as the capacity of suppliers to meet a project's needs and the wider programme's aggregate level of demand. It seeks to understand the exposure of suppliers in the context of their financial strength, the effect the programme has on this and also the wider economic impact of the programme. PSE therefore takes the view that the successful project is not as important as the successful programme, and that the successful programme is not achievable without the successful businesses that form its delivery supply chain.

Figure 1.1 below highlights the processes and strategies of the purchase engineering and supplier engineering streams that make up the PSE model. These streams are fundamentally linked to each other and assist in laying the foundations of the future success of a programme's delivery.

The purchase engineering and the supplier engineering functions together inform the procurement and delivery process and thereby ensure that the risks associated with procurement are managed, while opportunities are created and in doing so the client's value requirements are realised.

Purchase Engineering seeks to establish a strategy and methodology for assembling a purchasing 'machine' that can be used to procure any number of contracts efficiently to deliver the objectives of a client.

Similarly, Supplier Engineering aims to establish a framework for the programme supply chain team's interactions with the supply chain organisations that are seen as critical to the success of construction. This engagement is both during the pre-construction (procurement) and

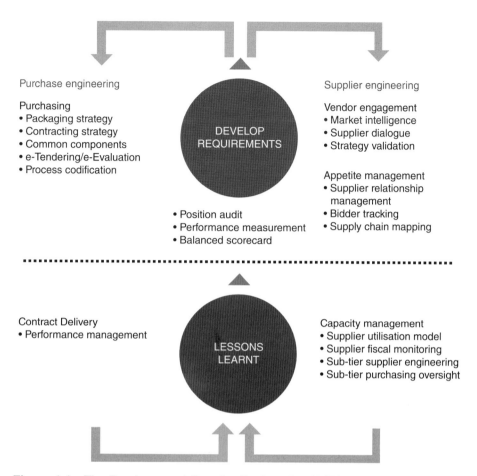

Figure 1.1 The Purchase and Supplier Engineering (PSE) model.

construction (delivery) phases. PSE helps to remove surprises during procurement and assists in the avoidance of certain project and programme supply chain risks during delivery that could otherwise lead to major cost and time overruns and undermine the value-for-money requirements of the client.

The PSE model is shown in Figure 1.1. The dotted line dividing the top and bottom halves of the diagram distinguishes between those elements of the model that are delivered during procurement, such as packaging strategy (above the line) and those that impact on delivery post-contract award, such as performance management (below the line).

Programme organisation – an Olympic case study

It is important to recognise at the outset that even large public-sector clients may be headed by people newly appointed to positions of authority. Often there is little qualified and experienced staff to support them and those that are available are not always to hand. This is partly the result of the practice of outsourcing expertise in the public sector. Moreover, there may be little build-up of relationships between the public sector and contractors over a prolonged period of time. By the nature of construction contracting, relationships tend to last only for the duration of projects; there is little opportunity for working relations to become established over a working life. This leads to problems of communication that need to be carefully managed.

For example, even before embarking on a project, contractors need to know what the client's requirements are. This goes beyond simply describing the building in technical terms. Large projects involve complex interactions, including social, environmental and economic aspects. Clients may initiate procurement in the built environment with an understanding of how they perceive success in a project, but without a clear way of explaining this to the supply chain.

Without a clear value definition, supply chains cannot respond to client requirements, as they do not necessarily understand their employers' expectations. Not being fully aware of these requirements, contractors cannot respond to these expectations clearly and coherently when putting tenders together. Clients also find it more difficult to evaluate tenders beyond the objective requirements of price and technical specifications, if they have not effectively communicated their value framework. The client might well have to meet demands from a variety of different stakeholders, with varying degrees of influence and power over their project, programme or portfolio.

These demands can be numerous and can come from inside the client organisation as well as from external stakeholders. The social and political implications need to be taken into account, as the impact of the addition of a new building on the built environment can affect third parties, such as local businesses, residents and amenities. Whether this impact is positive or negative, or is even considered, will depend on the client's priorities and their need or desire to take various factors into consideration. Publicly funded projects, for example, must demonstrate 'best value', but very often what represents value is unclear and sometimes even remains unstated beyond the most basic description. Conventionally, clients describe value in three relatively simple terms. These are cost, time and quality. However, that is no longer adequate or sufficient, as objectives extend beyond the built asset especially in the public sector, where there is often a need to take into account environmental factors, health and safety and the social impacts that may occur.

The Olympic Delivery Authority (ODA) was the public body put in place and made responsible for developing and building the new venues and infrastructure necessary for hosting the London 2012 Olympic and Paralympic Games. A delivery partner, CLM, was appointed via a public procurement competitive dialogue process to work with the ODA to programme-manage the delivery of the construction of the venues and infrastructure for the Games.

The ODA was established by the London Olympic Games and Paralympic Games Act (2006). The Act was passed to ensure that the necessary planning and preparation for the Games could take place. It allowed the ODA to:

- buy, sell and hold land;
- make arrangements for building works and develop transport and other infrastructure;
- develop a transport plan for the Games, with which other agencies had to cooperate, and make orders regulating traffic on the Olympic Road Network and Paralympic Route Network; and
- be the local planning authority for the Olympic Park area.

As a public body, the ODA was accountable to government, the Greater London Authority (GLA) and other stakeholders for its work.

As the client, the ODA decided it needed to mobilise resources from the private sector by first procuring a delivery partner (DP). The need for expediency and the requirement for a large number of experienced professionals to be mobilised quickly meant that, by buying a private-sector DP, the ODA could bring to bear the necessary resources to organise the procurement quickly. The DP approach also had other advantages in that it meant the ODA could deploy its resources flexibly, according to the programme's requirement at any given time. It also meant that the ODA could remain a 'thin' client, with comparatively low numbers of directly employed staff, who controlled and directed the experienced construction professionals working in the DP.

The competition to award the DP contract to work with the ODA was won by a joint venture organisation made up of three companies from different parts of the construction industry. The first was a large engineering organisation and programme manager, CH2M Hill; the second was a tier 1 contractor, Laing O'Rourke; and the third was a construction management organisation, Mace. These three organisations, along with a number of strategic partners (including Davis Langdon) became the DP, known as CLM.

The Delivery Partner was tasked with the day-to-day practicalities of managing and delivering the programme and administering the contracts for each of the construction projects. Moreover, among the members of the DP they held the necessary skill sets and, if the need arose, they could be the 'builder of last resort' and step in if required to assist in the construction process. It was the depth and breadth of the DP expertise that meant they were an extremely well-informed client delivery partner, with insights and a detailed understanding of the contractors' points of view.

In 2006, when the DP was mobilising to begin its work to deliver the programme, the global economy was confident and growing and the construction industry was experiencing high private-sector demand both at home and abroad, with this demand returning very good margins for contractors. However, widely publicised problems surrounding the late delivery of the Wembley Stadium project, a major sporting venue (eventually completed in 2007), had received much negative coverage, both at home and overseas. The criticism centred around the enormous overruns of both its budgets and schedules on what was agreed to be a complex stadium construction project. Therefore the prospect of a one-off government client procuring a number of sporting arenas and other associated infrastructure, which were to be built in a densely populated area of London, on highly contaminated land and to a schedule with an immoveable completion date (namely, the opening ceremony of the Olympic Games), meant that the tender opportunities offered by the Olympic Games infrastructure did not present an attractive business case to contractors when compared with alternative projects on offer from private-sector markets and their clients. These clients, after all, also had the added attraction of not being bound by the red tape of public contract regulations and EU procurement legislation.

Therefore, from the supply market's point of view, private-sector contracts offered not only greater flexibility to suppliers but – more importantly to large tier 1 contractors – the prospect of greater profits, repeat work and the potential for sustaining the growth of previous years. The appetite of contractors for the ODA's emerging project opportunities was therefore a major challenge confronting both the ODA and their delivery partner right from the outset. It appeared at best that contractors would be reluctant to enter into any dialogue with the ODA, and at worst would not even respond to any procurement opportunities associated with the Olympic programme. These were the circumstances that acted as the catalyst to produce the PSE approach.

Procurement organisation structure – the Olympic Delivery Authority

In 2006 the challenge was to set up a method of procurement that would meet all the strategic goals and surpass all the critical success factors required to deliver the ODA's value-for-money criteria. To achieve those ends, overall responsibility for the procurement function at the Olympic Delivery Authority (ODA) rested with the ODA Head of Procurement, to whom the delivery partner's (CLM) Head of Procurement reported. The DP's Head of Procurement managed two main functional streams: the programme procurement's strategic and operational teams. The programme supply chain management team was part of, and contributed to, both these functional streams. The procurement organisation within the ODA took the form of a classic matrix management structure, in which centralised specialist functions supported multiple individual projects. The matrix structure is commonly used in best-practice organisations, as it helps to maintain consistency and capture best practice, while allowing a high degree of flexibility. Figure 1.2 shows how the programme procurement-level specialist functions, strategy and governance, operations and support and supply chain management and assurance, supported the project-level procurements. In the case of programme supply chain management, their involvement went beyond procurement and influenced the delivery – hence the term 'Supplier Engineering'.

Roles and responsibilities

The roles and responsibilities for procurement staff were defined within detailed job descriptions. These job descriptions, as well as defining the

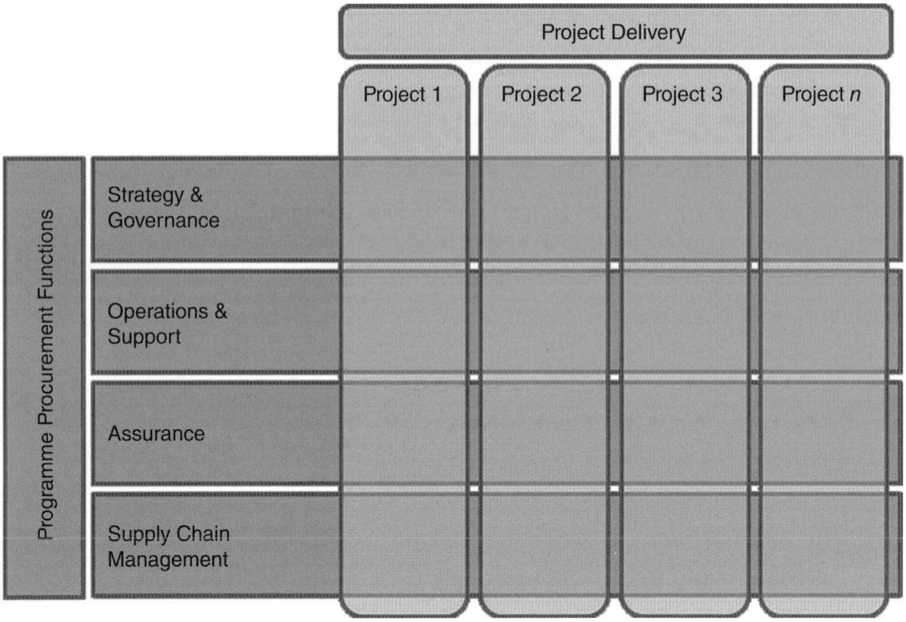

Figure 1.2 Programme-level procurement function matrix organisation.

roles' requirements, set out very clearly the obligations for all staff to comply with the ODA procurement code (the codified process for procurement). The ODA procurement code was developed as part of the PSE process and is described later in Chapter 7. The procurement code was based on similar procedures and working instructions from within other safety-critical industries, including rail and petrochemical. The purpose of the code was to make responsibilities during the buying process completely transparent and unambiguous. The codification also facilitated clear lines of reporting and communication. The benefits of an integrated team, consisting of the CLM Delivery Partner and the ODA project sponsors (who held responsibility for the project), included encouraging people to work together in order to generate the information required to satisfy the gateways within the process. The emphasis on team integration encouraged a more cooperative working environment and a greater sense of ownership within the project teams than might have been achieved in a more formalised client-and-supplier relationship.

Projects and programmes

Many of the tools and techniques described in this book are applied across a number of projects within a programme, to give an overarching view not necessarily available to any one firm actually working on one or more projects within a programme. For example, a main contractor may not be in a position to assess a subcontractor's or a supplier's exposure to the programme, because they are unaware of the competing commitments of other main contractors on the programme, who may be using the same suppliers.

At other times this could be viewed as interference in the business affairs of a supplier, but it is essential for monitoring the delivery of a programme of projects, where multiple projects are competing for the finite capacity of a specific supply pool. This programme approach to understanding demand ensures that firms are not overexposed on any one project and that, more importantly, the sum of exposure on all projects does not impact negatively on the programme as a whole. One successful project does not make a successful programme, and yet the very nature of project management is focused on delivering a successful project, often in ignorance and possibly to the detriment of other projects within a programme. Only when all projects within a programme have met their targets can programme managers claim to have satisfied the client's objectives and delivered success for the programme.

The principles described in this book provide the necessary guidance on establishing criteria to assess the success of a programme using a balanced scorecard approach. By establishing and communicating the targets of the programme, these form the criteria that influence the delivery of success for each of the individual projects that make up the programme. Most importantly, these criteria can also be used to guide clients in the selection of suitable suppliers in terms of their appetite, capacity and capability to deliver the programme's criteria for success.

While this book is concerned with the management of a programme of work for large, complex projects, the techniques can be applied to any programme or portfolio of projects. The same techniques can be used by contracting organisations, whose business is managing multiple projects. This approach is in contrast to the way firms operate in the construction industry, where projects are often viewed in isolation, as if they take place alone rather than in the context of the other projects running concurrently. Projects are often delivered in silos according to the needs of the individual client in question, and only passing attention is paid to any corporate strategic view – usually at head office, often with little translation of this on the project site. However, once a project starts on site, the project team are charged with delivering the project without any concern for the business as a whole or for any impact their decisions might have on other projects being delivered elsewhere. This is often true when projects are delivered by the same contractor, but for different clients. Changing this approach to projects, and taking a far more strategic approach to their procurement either at a programme or a project level, represents an opportunity for efficiency gains, risk avoidance and value improvement.

The aim of the book is to provide an understanding of the programme procurement principles developed and to give practical advice on avoiding many of the problems that can arise. For example, it is not sufficient to procure main contractors alone, while ignoring the impact of their selection of subcontractors and suppliers, because problems frequently arise in construction when subcontractors do not have the capacity and capability to service all their commitments on all their contracts. Labour and resources are often moved from one contract to another by subcontractors, depending on the demands of competing projects, the threat of penalty clauses or delayed payments for late delivery. This leverage over subcontractors is important, because resources are pulled by the leverage of their clients, who are often the main contractor. It should be recognised that the vast majority of main contractors are not vertically integrated and therefore they rely on their subcontractors to carry out the physical delivery of the construction process.

Concluding remarks

This chapter has introduced the concept and model of PSE in its most basic terms and has discussed some of the challenges the model was developed to address. It has also highlighted the way in which the programme procurement function works as a function within the wider programme management team and described how the ODA utilised third-party specialists in their delivery partner, CLM, to mobilise the appropriate resources for developing and implementing the necessary requirements for realising the ODA's vision as part of a fully integrated client function.

The next chapter introduces some of the economic theory and high-lights some of the market dynamics that act upon the construction industry. The success and growth of the construction sector often relies on the success and growth of the economy in general. The PSE approach was initially developed to respond to a market that was operating in an extremely buoyant global economy. When the recession began in 2007, during the delivery of the programme, the attention to detail and the systems that were developed to avoid the risks of a volatile marketplace enabled the teams working on the London 2012 Games to avoid the risks associated with the downturns in the economy, in particular the avoidance of supply chain failure. For this reason the authors feel it necessary to give some economic context, as it was this sensitivity to the economic environment that shaped the development of the approach now known as Purchase and Supplier Engineering (PSE).

Reference

Wolstenholme, A., (2009) *Never Waste a Good Crisis: A Review of Progress since Rethinking Construction and Thoughts for the Future*, London, Constructing Excellence. http://www.constructingexcellence.org.uk/pdf/Wolstenholme_Report_Oct_2009.pdf.

2

A framework for understanding markets in construction

External construction of the Velodrome almost completed (photo courtesy of Mark Lythaby).

Programme Procurement in Construction: Learning from London 2012, First Edition.
John M. Mead and Stephen Gruneberg.
© 2013 John Wiley & Sons, Ltd. Published 2013 by John Wiley & Sons, Ltd.

The in-situ cast concrete diving platforms of the Aquatics Centre (photo courtesy of Mark Lythaby).

Introduction

Undertaking a large programme at a time of economic crisis and managing it to a successful conclusion depended on both an understanding of the market and an appreciation of the risks involved. Managing the task called for careful preparation and an understanding of the market forces interacting with the programme. The previous chapter introduced Purchase and Supplier Engineering (PSE), which became the approach used to manage the procurement of the programme that oversaw the delivery of London 2012.

View of the Olympic Stadium from the North of the Olympic Park with crowds enjoying the action on the large screens (photo courtesy of AECOM).

This chapter gives an overview of the theoretical challenges facing both buying and supplying organisations, to give an insight into the thinking behind the development of PSE for delivering procurement on large construction programmes. From a general approach to managing the supply market, the chapter goes on to consider the relationship between the client and construction in general. This leads to a discussion on the distinction between projects and programmes and where the role of the client fits in with the supply chain. The chapter concludes with the issues of outsourcing and subcontracting and a discussion on understanding and managing conflict in construction.

Managing the supply market

In industries, where firms carry out several stages in the production process, from sourcing raw materials to delivering finished goods to customers, the organisation of production is hierarchical. Departmental managers within the same firm collaborate through rules, procedures and levels of authority (or hierarchies) to ensure a steady production flow. This is very much in contrast to the way construction production is organised and managed on site. Because the construction industry is fragmented into many specialisms and disparate organisations, it can be characterised as market-driven. Because trading organisations in the construction industry are by definition not vertically integrated, they are forced to find and buy the services of other firms in order to deliver their part of the project.

Firms in construction therefore rely on transactions with other firms to obtain the necessary skills, materials and plant they require, as and when they need them. Transactions involve buyers on the demand side and sellers on the supply side. In its simplest form the interaction between supply and demand allows the two sides to a transaction to arrive at a price both parties feel is acceptable, given the constraints set by the other party or parties to the deal.

A simple textbook analysis of supply and demand is sufficient to understand the challenges faced by the procurement team of the London 2012 Olympic Games, or any other major capital works programme. Starting with the basics: according to the first law of demand and assuming all else remains the same, the higher the price the less will be demanded, and the lower the price the more will be demanded. This is shown in Figure 2.1, where price is on the vertical axis and quantity is shown on the horizontal axis. Assume firms compete for a given project or amount of work available. This is the amount demanded and is represented by the vertical demand line. Regardless of the price, the construction work on offer remains the same. As price decreases from P_2 to P_1, the quantity demanded remains at Q_1.

However, according to the law of supply, the higher the price the more will be supplied, and the lower the price, the less will be supplied –

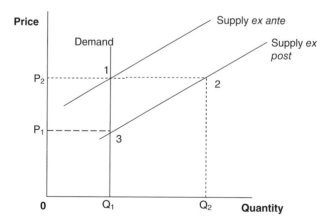

Figure 2.1   The manipulation of supply.

assuming, of course, that everything else remains unchanged. The market price occurs where the demand line and supply line intersect, which matches the quantity on offer with the quantity demanded. Prior to any engagement with the supply market, this occurs at intersection 1 in Figure 2.2, where the demand for construction meets the 'supply *ex ante*' line.

Each supply line represents a given set of circumstances and conditions, including a given number of potential tenderers and their level of interest in participating in a project. The aim of the procurement team is to increase both the pool of firms wishing to compete for the work, and their appetite for the work. Achieving these objectives shifts the whole supply curve to the right, which implies that, at the same price, more firms are willing and able to offer their services to the client. This occurs at point 2 in Figure 2.1, which identifies the new hypothetical quantity all firms in total would be willing and able to offer for sale to the client – an increase from Q_1 to Q_2. An increase of supply would occur at every price, if the procurement team is successful in its efforts to attract contractors and suppliers. The aim of the procurement team is therefore to shift the supply line from the position before the client engages with the supply market to the supply line *ex post* engagement.

As a result, the client now has more firms competing for the same amount of work and a surplus or excess supply has been created. The surplus capacity now on offer puts a downward pressure on price to point 3 in Figure 2.1. By increasing the appetite of firms to participate in a project, the client acquires the services of contractors at a lower price than would otherwise have occurred.

Although this basic approach fits well with the tactics of procurement engagement strategy, a number of non-theoretical issues are raised. The client has a given project and requirement and the quantity of services cannot be varied easily. In economic parlance, demand is said to be inelastic or relatively invariable, regardless of price.

Moreover, the actual process of bidding for work is not a marketplace as much as an auction room, with one buyer and several suppliers competing on, among other things, price to win the tender. Often this leads to a concentration on price, and the lowest tender price is frequently the winning criterion. On major projects such as the Olympic Games and Crossrail, where public interest is also of major importance, other selection criteria need to be considered, and both of these programmes applied greater weight to quality elements than they did to pricing alone.

Non-price competition can relate to a number of factors, such as reliability, reputation for punctual delivery, and quality of output. In later chapters these factors are discussed in more detail.

A further aspect of the market orientation of the construction industry is the part played by transaction costs. In his seminal work on transaction costs, Williamson (1975) identified a number of business practices that added to the cost of transactions over and above the price paid. These included opportunism, uncertainty and bounded rationality, which relates to the need to make business decisions with only limited time and resources to acquire the necessary information. Other transaction costs arise out of what Williamson refers to as 'information impactedness', which refers to the situation in which one firm has technical or commercial knowledge that the other party does not possess, so the party with the knowledge can take advantage of the other.

Although transaction costs are often assumed not to occur within single organisations, competition between departments can lead to transaction costs even where production is hierarchical. The existence of transaction costs has been recognised, but their measurement is elusive. It is simply not possible to assess the value of transaction costs in construction, but one of the most important transaction costs in construction arises out of the tendering process, whereby firms invest heavily in several tender competitions, to win only some of them. The cost of tendering is borne by contractors, who must cover the cost of their failed bids out of those contracts they actually win. This is therefore another hidden transaction cost.

Transaction costs are not restricted to construction. In most industries the production process is similarly fragmented. Outside suppliers provide firms with components, which they then assemble. The construction industry is no different. However, the suppliers and the types of firm vary between industries. Suppliers in the construction industry can be classified as consultants, main contractors, trade contractors and manufacturers and material suppliers.

Consultants provide design, engineering, commercial and project management services. Clients generally enter into a contract with a main contractor, who in turn subcontracts work to different trade contractors. They in turn procure the suppliers and manufacturers, who provide the products and goods. This chain of transactions often means that the client can be contractually quite far removed from the organisations that ultimately provide the goods and services used to produce the end construction product, and the lines of communication can be quite extended.

The network of contracts between all participants in the construction process formally defines how a project, however complex, is put together. For example, in the tendering process contractors compete on price by calculating their production costs, management overheads, resources used and a mark-up, taking their allocation of risk into account. Contractors manage their exposure to risk and uncertainty by parcelling the work into numerous subcontract packages. This implies that a core competency for contractors is the management of subcontractor prices. As a result, there is a disengagement from the actual building process by main contractors.

Each transaction in the supply chain represents an agreement to purchase a supply of a service or a product. Each transaction takes place within a particular market, with its own unique set of characteristics. For example, in some markets there may be many variants from which to select. Each manufacturer produces a differentiated product or service, with features only their offer can supply. Nevertheless, competing offers may still be easily substituted for that of a competitor.

Other markets involve a service that is assumed to be identical regardless of the supplier, such as the general contracting market. Some markets require technical knowledge and skills; others require specialist designs. The degree of competition between suppliers varies depending on the ease with which new firms can enter. Where the barriers to entry are low, the market is described as contestable. In contestable markets, even though there may not be many competing firms, the prices offered tend to be similar to competitive prices, because firms in the market fear attracting new entrants, who might reduce prices even further. Even where there are barriers to entry because of the expense of plant and machinery, competition can be fierce in a buyers' market and less so in a sellers' market. A buyers' market occurs when suppliers find it difficult to attract customers and offer discounts and additional services to win orders. A sellers' market arises, when firms operate at or near their full capacity and have to be tempted by clients through higher prices, prompt payments or even payments in advance. Because of the fragmented nature of construction, it is possible for buyers' markets to coexist alongside sellers' markets on the same project.

The client and construction

With the exception of the very smallest building projects, the construction process is invariably complex, involving several different construction firms and a number of often diverse material, plant and labour providers. This is due to the large variety of materials, components, plant and machinery, knowledge and skills needed to come together to complete a construction project.

From the clients' perspective, Boyd and Chinyio (2006) list the failures of the industry as poor value for money and poor client relationships, as well as poor quality, late delivery and overrunning budgets. Engaging with the construction industry is often seen as difficult, adversarial and litigious. However, it is precisely because of those aspects of the construction industry that clients have a responsibility to understand, influence and perhaps even control certain aspects of the building projects and programmes they undertake. It is not sufficient to expect the industry to deliver simply by placing an order with a main contractor, based on a tender bidding process.

Clients need to engage fully with the process. Rimmer (2009), for example, sees the client as a key player in the process. Clients need to see procurement as a continuous process comprising many stages, from inception, through tendering and construction to completion and beyond. There is simply no structure in place that does this for the client from within the industry – not even the concept of the main contractor. Risk and responsibility are often passed on by main contractors to subcontractors and at this point the client's visibility of the management of the risk disappears, their ability to influence diminishes and the strategic goals are lost.

Boyd and Chinyio (2006) distinguish the strategic aims of organisations in the private and public sectors. While firms in the private sector are concerned to achieve a number of priorities, including profits, business efficiency, productivity and market position, the strategic aims of public-sector organisations focus on accountability, transparency, social involvement, equity, employee welfare and effective service. Procedures in the public sector therefore tend to be structured and bureaucratic (often for reasons of accountability). These procedures need to take into account the priorities of politicians, the resources needed and the range of stakeholders affected. Hence, the management of construction in the public sector includes the production of documents that present procedures and criteria consistent with stated political objectives.

In their approach, Boyd and Chinyio (ibid.) see the contractor as the initiator of change in the client. While that may often be necessary, it is our contention that by its very nature, a contractor mobilises resources only in response to client demand. Only when a client approaches a contractor, or engages with the industry through invitations to tender, do contractors respond. As they do not initiate construction projects, contractors are passive players in the construction process, carrying out only what they have been asked to do by developers in both the private and public sectors. It is therefore incumbent on major clients to take responsibility for the whole construction process, from inception through to completion. That is what is meant in this book by construction procurement.

Projects, programmes and construction dynamics

Do not leave anything to chance. To avoid surprise, foresight, planning and information are key issues. As a client with a major capital investment project to commission, it is not acceptable to make assumptions. Knowledge and awareness are vital. Control of the situation is essential. For those reasons, the client team must remain fully informed and involved with the construction process at all stages, in order to take decisions and lead. Otherwise, leadership falls to another organisation, such as the contractor, the architect or the consultant, and each will have an agenda that differs from the client's and will act in their own interests at the expense of the client.

Nevertheless, Wolstenholme (2009) predicted that clients would no longer lead change in the construction industry, and it was up to contractors to innovate to improve the construction process through collaboration and integrated working. While it is perfectly possible for innovation to come from within the construction industry, it tends to be driven by informed client leadership and directed by their demand. The supply chain simply does not possess the strategic means or ability to control the demand that creates opportunities to take the lead, owing to both its fragmentation and its market-orientated outsourcing methods. Necessary and timely information is not always forthcoming and, when it is, it is not necessarily pooled in a process of shared values and common interest.

Hagan, Bower and Smith (2011) recognised that within any project-based firm, projects can be interrelated in terms of their use of resources. Hagan *et al.* aimed to find an appropriate method for managing different projects within a multi-project contractor. The need for this arises because of poor performance by the construction industry. According to Wolstenholme (2009), there is resistance to change. Baccarini (1996) and

Remington and Pollack (2007) cite the complexity of projects and managing the supply chain in creating situations in which problems are inevitable. Egan (1998) found that the lack of attention to client needs leads to disappointment in the building process. However it has been noted, by Artto and Dietrich (2004), Danilovic and Borjesson (2001) and Payne (1995), that many of those involved in the construction and development process do not themselves understand the problems that arise in managing projects alongside other projects. Few are ever given the opportunity to gain an overview of the aggregate effect of simultaneous and conflicting projects competing for limited resources. There are numerous aspects of complexity that need to be taken into account and understood. This book is concerned with practical approaches for dealing with the complexities that arise at different stages in the delivery of large programmes. However, the techniques and methods described here can be applied equally well across a client's portfolio of projects, or within a business that manages multiple projects and also, to varying but lesser degrees, on individual and smaller projects.

The key difference between a client responsible for a very large programme, as described in this book, compared with an individual contractor forming part of the supply chain on a programme, is that the client team is in a position to gain an overview of their total demand and thus a large proportion of a market in a given industry, sector or location. This means that the client team can observe what competing contractors and their supply chains are doing in a way that individual contractors cannot see, because tier 1 contractors (main contractors) are in competition with the other tier 1 contractors across the programme. Thus vital information is always denied to main contractors for reasons of market dynamics, commercial confidentiality and competitive secrecy. The client commissioning a large programme or portfolio of projects, however, is in a position of necessity and needs to know the capacity of potential tier 1 and tier 2 contractors (subcontractors) before work can be let. This gives the client an opportunity to see what overlaps exist, what clashes in timing might be avoided and what capacity each contractor can draw upon to meet its obligations. Only the client can gain this more strategic and holistic view of the micro-market they create through the scale of their demand. The supply side is in competition with itself, and both tier 1 and tier 2 contractors need to form strategic alliances in certain areas to secure exclusive specialist input to their offer and therefore they cannot share information – which in any case could be seen as anticompetitive collusion in terms of the Competition Act, 1998 in the UK.

Nevertheless, project management is often seen in terms of a single project carried out by one firm and its supply chain, assuming the project will be barely affected by the existence of other projects. This is an oversimplification, according to Sacks (2004), Hossain and Ruwanpura (2008) and Pellegrinelli (2011), and runs in the face of reality, where many projects impact on each other in an uncoordinated, unplanned way, leading to unanticipated difficulties. Antoniadis, Edum-Fotwe and Thorpe (2011) pointed out the need to understand the nature of the various connections between all construction firms, including those in the vicinity of the project but not directly involved. Firms working on other projects in the area may affect progress and the capacity of the local construction supply market to meet demand.

A project must be viewed in the context of the supply market, which is comprised of potential tier 1 and tier 2 contractors and many other layers of specialist construction firms, as well as component and material suppliers, each with their own expertise and skills to offer in an ever-changing construction market. It can be seen as a system consisting of organisations operating in a business environment, in which multi-project firms struggle to allocate resources between competing projects, and that Engwall and Jerbrant (2003) refer to as the 'resource allocation syndrome'. This defines the framework of systems theory, which identifies the concepts needed to understand and manage the procurement and production of the built product. In one systems-theory approach there are six interdependent and interacting concepts; they are: decision making, goals, the product, resources, people and the process. A full discussion of systems theory lies beyond the scope of this book, but it is this systems approach that creates order in what would otherwise appear to be a complex, chaotic, unstructured, unpredictable and ever-changing situation. Instead it simplifies, defines and prioritises the issues that need to be dealt with. This theoretical approach lies behind the practical applications used to solve the management issues as and when they arise at each stage in the procurement process.

The client and the supply chain

Even modest construction projects involve large investments and risk. When construction projects as large as the infrastructure for the London 2012 Olympics are undertaken, they comprise multiple and diverse large and small projects, which together form a programme. As a client, it is not always efficient and neither is it sufficient to pass the responsibility for projects or programmes on to main contractors – often referred to as tier 1 contractors – and remain at arm's length from the process.

Of course, this implies that the client must engage with the process without compromising the contractual relations between the client and the main contractor. This is a very delicate issue, but the methods and tools described in later chapters demonstrate how it was achieved for the delivery of the stadia and infrastructure for the London 2012 Olympics. Partly as a result of the techniques described in this book and used during procurement and delivery, the programme was delivered on time and under budget.

Based on data taken from the *Construction Statistics Annual* (ONS, 2011), between 1997 and 2006 approximately 22 per cent of all construction work was public-sector construction, and between 2006 to 2010 this figure rose to 25 per cent on average, with a peak of 31 per cent in the final quarter of 2009 – in part due to work on the 2012 Olympic Games. In both the private and public sectors, clients who commission most large and complex one-off projects or programmes rarely have sufficient in-house skills to manage the entire programme of work.

The nature of demand for a one-off programme requires a client to be able to mobilise its resources quickly and efficiently. This was difficult for the ODA on the Olympics, as they were seeking to mobilise staff during an economic boom, when many construction experts were already busy delivering more profitable private-sector projects and programmes than those offered in the public sector. It was for this reason that a Delivery Partner model was chosen, as it offered the client the ability to draw upon expert resources as demand dictated during delivery, offering both commercial flexibility and agility of resource supply. This was especially important in the initial mobilisation, when large amounts of professional expertise were required to develop and implement processes and systems and generally get things 'up and running'. To meet their delivery deadline, the ODA in effect had to go from being an organisation with a fairly modest annual turnover to one with a turnover in excess of £1 billion per year, within a little over a year of being formed.

Defining the supply chain

The construction process relies on many different specialist firms working together as subcontractors to a tier 1 main contractor, who acts

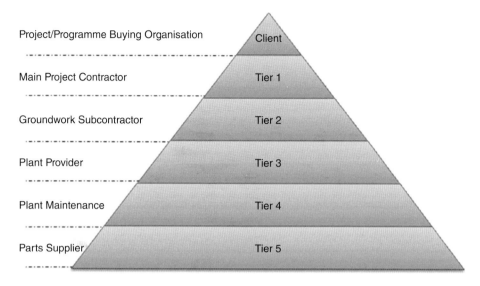

Project/Programme Buying Organisation — Client

Main Project Contractor — Tier 1

Groundwork Subcontractor — Tier 2

Plant Provider — Tier 3

Plant Maintenance — Tier 4

Parts Supplier — Tier 5

Figure 2.2   Example of a supply chain hierarchy.

as integrator. These specialist firms are tier 2 suppliers and they in turn may subcontract or be supplied by other firms, namely, the tier 3 contractors. Taken together, the tiers of organisations comprise the supply chain. Figure 2.2 illustrates the supply chain hierarchy, beginning with the client or buying organisation. The client creates an opportunity for tier 1 contractors, who may be major contractors with the financial strength and capability to undertake the management of the project. They assume overall responsibility for the delivery of a project or programme element. The tier 1 contractor then subcontracts to tier 2 specialist firms, who possess the required skills and experience and are of sufficient size to accept responsibility for their specialist area, with the knowledge and contacts to further sublet certain elements of their work to tier 3 contractors, who have the expertise and skills to carry out the work. These lower-tier firms are often small or medium-sized enterprises (SME), who in turn may further subcontract to others further down the supply chain. The number of firms at each tier increases until as many as 40 to 80 firms may be involved on any one project, depending on its size and the nature of the tier 2 works, as some supply chains are more technically complex than others. In this way the delivery, capability and financial capacity risks are spread over many participating firms, while accessing the specialist labour, plant and materials required for the works.

Figure 2.2 gives an example of a single supply chain from client to tier 5. Of course, the supply chain does not necessarily end at tier 5, but could go beyond that as in the example above to the plant parts manufacturer at tier 6 and then on to the parts manufacturer's own supply chain. The purpose of the diagram is to highlight the complexity, interdependency and diversity of the supply chain.

A private-sector perspective of supply chains is given by Rimmer (2009), who defines the management of the supply chain as the 'management of upstream and downstream relationships with clients and suppliers to achieve greater project value at less cost' (Rimmer, p. 153). He defines value and cost from the client's point of view. Value is seen as any aspect of a project of importance to the client, while cost is the total amount paid by the client, including interest charges (Rimmer, pp. 138–9). To Rimmer the purpose of supply chain management (SCM) includes the delivery of value, the elimination of waste and the measurement of performance.

Although there are many exceptions, much of the literature on the management of construction projects and their supply chains is viewed from the main contractor's point of view. Even Rimmer's description of SCM can be seen as referring to upstream and downstream relationships from the point of view of the main contractor. It is assumed contracts between buying organisations and tier 1 contractors pass the associated supply chain risk of their subcontractors to the tier 1 contractors, and it is therefore their responsibility to manage those risks. Such arrangements often work well on single contracts or one-off projects and sometimes even on portfolios of projects, where the same tier 1 contractor is engaged to deliver multiple projects for the same client concurrently. The problem with this approach is that it ignores the role of the client and their indirect relationship with the tier 2, tier 3 and tier 4 contractors, often with severe and avoidable consequences affecting the delivery of a programme.

A large programme of projects over an extended period, using a number of tier 1 contractors on different projects, raises particular management and coordination issues. These require meticulous management of detail, without losing sight of the large scale of the whole programme and the priorities and aims of the client. Maintaining this balance is the challenge faced by the client delivery team when managing large construction programmes. Later chapters will provide the tools for dealing with the issues involved. However, many of the tools and techniques described and the lessons learned can also be applied to individual projects of any size, where a main contractor engages a supply chain consisting of a number of specialist firms to carry out the work.

The organisation of a construction programme is a complex arrangement of businesses that form a team to deliver multiple contracts. Often these contracts differ from tier 1 to tier 2; for example, a tier 1 contractor may be on an actual cost contract, while the tier 2 contractors are more likely to be on a lump-sum type of subcontract. The team is headed by the client body, which is responsible for deciding how the project will be delivered.

Outsourcing and subcontracting

In virtually every industry, the outsourcing of non-core activities is widespread. For example, McCarthy and Anagnostou (2004) view outsourcing as a common practice in manufacturing, where firms often outsource a number of activities, including logistics, IT, accounting functions and legal services. Where firms carry out upstream functions they would otherwise have to purchase from lower tiers of the supply chain, economists refer to it as vertical integration. Vertically integrated firms are the exception in construction, owing to fragmentation and specialisation even by main contractors. Much of the production process is seen by specialist main contractors as outside their core activity, and therefore the proportion of construction output that is outsourced in the building industry by main contractors is very high.

Because there is perhaps less vertical integration in construction than in most other industries, clients need to gain an appreciation of the extent of outsourcing and the widespread use of subcontracting of the many specialist firms in the building industry. Clients should therefore seek to have visibility and influence on these important aspects of the value chain for delivery. The lack of vertical integration of firms has led to the fragmentation of the construction production process. As a result, many different and independent firms come together only at the final place of a product's delivery: the construction site, which is an environment of ever-changing and uncontrollable conditions that add to the variables of multi-organisational and multi-disciplinary interface delivery. This disparate group of specialist firms is a significant part of the supply chain, which also includes the material and component manufacturers and suppliers, who do not necessarily visit the building site.

Later chapters will outline the PSE strategies and techniques that have been used on the largest construction programmes in the UK delivered in modern times – namely, London 2012 and Crossrail – to ensure that projects are procured well and clients offered supply chain visibility and influence in order that they can be delivered as planned and realise value. Of course, there can never be certainty in construction and even if the advice given in later chapters is followed, it will not necessarily produce the desired result – but it will help to reduce the number and severity of adverse surprises and improve value, however defined.

Complex projects are unpredictable, unique and dependent on many factors in the wider economy beyond the control of the management of construction programmes. However, this book highlights the innovative methods that were adopted in the delivery of the procurement and management of the supply chain of the London 2012 Olympic Games and the Crossrail programme in London. These were programmes with a total value well in excess of £20 billion and made up of hundreds of construction contracts valued between several millions of pounds to hundreds of millions, which in turn led to thousands of supply chain subcontracts.

'Over 1600 companies, 98 per cent of them registered in the United Kingdom, were our tier 1 contractors, with contracts directly with the ODA, with thousands more down the supply chain' (Armitt, 2012).

Risks on small projects can be minor irritants that cause delay and relatively small budget or schedule overruns. However, when equivalent irritants occur on large programmes, they can be compounded and lead to major impacts not only on the project concerned, but also on the programme, and can even have an impact on the micro-economy of the region. The spill-over effects can affect firms across multiple contracts, related and unrelated construction projects and the overall programme delivery. Removing these risks can improve the chances of delivery targets being maintained and may even bring about schedule and budgetary savings. These knock-on effects can therefore escalate in value, with an impact far greater than the cost of the original cause.

Construction programmes have economic impacts on the wider community beyond the programmes themselves. The investment expenditure earned by many different firms in a programme's supply chain circulates around the economy, as additional employment is created that would not have come about without the new construction work taking place. This can be seen as the direct impact of the investment. Those who receive the additional income generated in turn increase their spending on goods and services quite separate from the construction programme.

The programme can be seen as having a multiplier effect on output in general, as demand increases with further increases in employment. As a result, incomes increase and subsequently spending rises, which has the effect of raising value-added tax (VAT) returns to the government, and drawing in imports. These expenditures can be seen as the indirect effects of investment. This is, of course, known as the multiplier effect. Similar effects have been noted and assessed by Kim, Park and Lee (2010), for example as a consequence of building a major convention centre.

In a recent report, LEK Consulting (2012) repeated an earlier calculation that the construction multiplier effect was as large as 2.84. This implies that for every £1m spent, the end result is an increase in total income of £2.84m. Others, such as Barro (2009), find this argument spurious, as it implies that the more the government spends, the more the economy gains. They take the view that the multiplier effect is equal to, or less than, 1. They argue that investments cannot be justified in terms of the multiplier effect, but only in terms of the long-term benefits that result from an investment over its lifetime.

Chapter 4 of this book describes how an overall programme procurement strategy is assembled and how that strategy may be translated into the delivery of a built structure. More specifically the chapter will describe the components of a programme procurement strategy that includes a packaging strategy, a contracting strategy and a supply chain management strategy. In describing these three sub-strategies, we extend the conventional practice of construction procurement and take into account many factors that are usually overlooked, and yet repeatedly lead to programme and project failure – late delivery, cost overruns or poor-quality delivery requiring remedial work. For example, we present a method for testing the capacity of tier 2 and tier 3 contractors to take on the work for which they are bidding, in order to anticipate the impact that might have on their ability to supply and the problems associated with it.

We describe how clients need to retain an appropriate level of visibility and control over their programme, even after contracts have been let and risks have been assigned. This is essential as certain risks – for example, time – can never be fully transferred to contractors. Even with the most onerous liquidated and ascertained damages (LADs) in place, for programmes such as the Olympics time was of the essence, as the opening ceremony was scheduled to take place on 27 July 2012, whatever happened, and this risk ultimately remained with the ODA and the UK Government. No contract could transfer the risk or compensate for late delivery.

The tools described here provide a level of visibility and transparency to ensure that directly contracted suppliers to the client apply similar levels of due diligence during their supply chain procurement to those applied in their own appointment and delivery. This empowers clients or their agents to make appropriate interventions in a timely manner, when performance standards are predicted to suffer or are not met.

To use a football analogy, if the contractors and their supply chains are the players on the field and the client is the club owner, then the manager of the team is the programme manager. Within this management team are the talent scouts and fitness coaches. This book describes the work of these specialist backroom staff, which identifies the best players to deliver, the managers required, playing formations and the required skills, and together they produce the results on the field of play. It is the fitness coaches who assess their players' fitness to play at their best and therefore ensure that they are able to deliver the manager's results and the client's requirement for silverware.

Understanding and managing conflict in construction

In his report on the construction industry called *Rethinking Construction*, Egan (1998) argued in favour of integrated teams and a quality-driven agenda to improve the performance of the construction industry. Although the report was about working practices in the construction industry, it did not suggest how integrated teams and quality agendas could be achieved in practice on site. Nevertheless, the report did encourage partnering arrangements and supply chain integration.

Kumaraswamy and Rahman (2006) also support the view that somehow outstanding performance in the construction industry can be achieved only by capable, integrated and motivated teams. While that is clearly the case, it is far from obvious how it may be achieved. Kumaraswamy and Rahman claim that it is not difficult to create an integrated building team, but accept that it requires a major cultural shift and a number of management tools. We disagree that the task of building an integrated team is easy, because we think this approach ignores many problems that arise in the course of assembling a construction team.

Turning a blind eye to the problems that arise, Kumaraswamy and Rahman (2006) refer to relational contracting based on common objectives and collaborative approaches, designed to reduce friction and disputes in the construction team. Moreover, partnering and alliancing arrangements were introduced in the 1980s, based on win–win principles. It is important to recognise the reality of working relationships in construction projects to manage the risks without relying on wishful thinking.

For example, in any business transaction there is an inherent conflict between the purchaser and the seller, and this conflict cannot be ignored. It stems from the differences in the aspiration of the buyer to reduce costs, while the seller seeks to achieve a relatively high return. The aims of each party can be achieved only to a greater or lesser extent, depending on their individual negotiating skills and their power and leverage in the market. Cox (2009) explains that there are three possible outcomes for each party, based on fully achieving, partially achieving or not achieving their desired outcome. In game theory the various outcomes of games are referred to as zero-sum, nonzero-sum and negative sum. In a zero-sum game, the winner wins at the expense of the loser. In a nonzero-sum game both players stand to win, and in a negative-sum game both stand to lose. As most games are either zero-sum or negative-sum games, harmonious, collaborative working in integrated groups in construction projects is extremely rare.

Cox and Ireland (2002) explain the relationship between main contractors and their subcontractors in terms of their roles as buyers and suppliers, a relationship that depends on the relative subjective utility and scarcity of the resources being exchanged between them. Each firm's understanding of its own and the other side's attitudes towards the resources and services being exchanged enables it to assess its power over the other party. Power rests on a number of attributes, including the number of buyers relative to the number of suppliers, the proportion of the supplier's capacity used by the buyer, and the cost of changing supplier.

Although Cox (2004) has pointed out the role of power in the interplay between main contractors and subcontractors, the supply chain management literature has often referred to the client as a key participant in setting up supply chains. However, King and Pitt (2009) point out that only a few clients have the size and repeat business to offer main contractors the incentive to collaborate with their supply chain, so as to allow their employers to determine the nature of their relationships with their subcontractors and the supply chain.

It is therefore not surprising that Dainty, Briscoe and Millet (2001) found mistrust and scepticism amongst subcontractors and went on to suggest that leading clients need to take the initiative if this barrier to improved working is to be overcome. That is precisely what occurred on the London 2012 programme, and this book describes the techniques and tools that were used to obtain improvements in the supply chain. King and Pitt (2009) state that supply chain management needs to take into account specific issues surrounding the industry, the client, suppliers and organisational factors. More specifically, they suggest developing the supply chain through the use of meetings, seminars and internal key performance indicators, with a view to developing trust, commitment, integrity and flexibility in the supply chain. They also see certain cultural values as necessary elements in a supply chain. These include a strategy for continuous improvement, health and safety standards and the early involvement of subcontractors. If these outcomes for the tier 1 and 2 firms in the supply chain are to be achieved, they require a number of detailed techniques, which are brought together by the PSE approach described in this book.

Simply winning at the expense of the other party (as in a zero-sum game) is not always the most desirable outcome, even for the winner. Depriving a supplier of profits may mean that the supplier drops out of the supply chain to the detriment of the programme, and means that it ceases to compete with other potential suppliers in the future. This increases the market power of the remaining suppliers in future tenders for the same client. Nevertheless, Cox (2009) argues that the majority of transactions involve zero-sum games with risks for both sides. As a result it is important to recognise that the management of projects is not about ignoring differences between clients and tier 1 contractors, or between tier 1 contractors and tier 2 contractors, but managing risks to avoid the most severe adverse consequences for all parties. It is therefore necessary to recognise the different interests of firms in the construction process and to accommodate these differences in management processes, as described in later chapters.

Other literature focuses on project management as a purely construction-phase, on-site activity, such as the Code of Practice for

Project Management published by the CIOB (2009). Success in project and programme management is often measured in terms of cost and timely delivery. For example, the National Audit Office (NAO, 2003) compared traditionally procured projects for the public sector with those procured using the private finance initiative (PFI). According to the NAO, almost 75 per cent of construction work using traditional procurement methods exceeded the agreed contract price and 70 per cent involved late completion, compared with 22 per cent and 24 per cent, respectively, using PFI. However, many other criteria – often qualitative in nature – also contribute to programme success, especially as a large number of diverse stakeholders with conflicting interests all have to be served when undertaking large public-sector construction projects.

In his report *Rethinking Construction*, Egan (1998) also advocated 'effective measurement of performance' as an essential prerequisite for managing and improving performance. This was reinforced by the NAO (2001), who suggested that failure to define requirements, poor briefing and inadequate quality assurance all inhibited construction performance. In a later report, the NAO (2005) advocated investing time and resources in the early planning phase of construction, clear communication from the very beginning of the tender evaluation process giving relative weightings to the different criteria to be used, and improving measures of performance.

In later chapters we describe both the long and deliberate lead-in period required for programme preparation and the proper full monitoring processes required to execute the plans to reach a satisfactory completion of a programme. Nothing can guarantee success, of course, but the fundamental challenge of major construction projects is to be focused on the overall aims and objectives while still dealing with the many thousands of details and individual decisions that need to be taken. Nothing of any significance can be left to chance. Risk has to be identified, and then it can be managed.

As Loosemore (2006) states, the fundamental key to risk management is the need to communicate, consult and involve all firms in the supply chain in the decision-making process to a greater or lesser extent. In terms of the organisation of the construction process, this may be achieved through supply chain management. There are two types of supply chain interaction: based on projects or on organisations. Project supply chains describe the supply chain involved in a particular project for a specific client. Organisational supply chains are established by main contractors, who may use the same specialist contractors on a number of projects, developing ongoing relationships with a number of suppliers, forming a virtual organisation capable of repeatedly collaborating on different projects for different clients. This may come about owing to the main contractor's ability to channel work and provide continuity. At the same time, familiarity with personnel and working practices can enhance trust between firms and produce productivity gains. However, the relationship remains essentially a set of discrete transactions and main contractors should still want to market-test their suppliers to compare prices and ensure competitiveness in the tender process.

Organisational supply chains in construction can best be described as a number of transactions between independent firms in the supply chain, and the relationships between the tier 1, tier 2 and tier 3 contractors are a form of economic game. Cox (2009), for example, distinguishes between one-off games and repeat games. Most games in construction are one-off, with neither firm expecting to continue the relationship on completion of the transaction. However, where more work is expected the transaction can take the form of a repeat game, in which both parties stand to gain in future transactions, if their relationship is profitable.

Cox also distinguishes the level of involvement firms may have with each other: namely, proactive or reactive, arm's length or collaborative. In fact, Cox (ibid.) goes on to develop a number of approaches to sourcing suppliers available to tier 1 contractors. These include sourcing from within the tier 1 contractor itself. However, the main contractor simply does not often have the in-house capability to undertake the work alone. A second method involves working closely with a tier 2 supplier, developing a long-term relationship and commitment, often based on the need to invest in plant and machinery by both parties to fulfil the requirements of a programme. Supply chain management simply takes this one step further and involves a close working relationship with a number of tier 2 and even tier 3 suppliers. Although many construction clients and some developers do not have the skills to organise their supply chains, supply chain management may be an appropriate approach in certain circumstances, where the client has a regular flow of standard construction work or where there is a dominant purchaser and the programme is on a large scale.

In SCM, clients need to be willing to engage directly with contractors and specialist firms as early as possible. However in traditional contracting, architects and engineers can often form barriers between the client and the contractor. Therefore, selecting an appropriate procurement route from among traditional, construction management, management contracting or design-and-build is necessary to reduce the barriers that impede the use of SCM.

Concluding remarks

PSE applies much of the economic theory highlighted and the issues identified in this chapter, by seeking to balance supply and demand. Through anticipating impending demand, the supply side is in a strong position to plan and mobilise resources to meet that demand efficiently. Timing and communication are the keys to this approach. Allowing supply enough time to respond, and ensuring that demand does not completely dictate the parameters for supply, helps to match needs to capacity, thus enabling an amicable and engaged supply chain to meet a well-measured and communicated demand.

The following chapters set out the PSE management model that has been delivered on the two largest construction programmes seen in the UK in recent years. Individually they represent a mobilisation of resources not seen in the UK since the Second World War. That is not to say that Purchase and Supplier Engineering (PSE) is applicable only to large-scale construction programmes or projects; it applies equally to smaller programmes and portfolios of projects. This book aims to inform the industry of a new way of thinking and approaching construction. The approach is equally applicable to construction clients and tier 1 contractors, as they are both buyers of supply chains for multiple projects, and they both rely on successful businesses in the supply chain to deliver value.

However, this book does not claim to solve all problems and neither is it the sole source of success on large programmes. It represents a starting point and offers a complementary process to established project management, contract administration, cost management and other traditional construction management techniques. Neither does the book advocate a completely new discipline. Procurement and supply chain management should still form part of any project or programme delivery. PSE only takes procurement and SCM to a more strategic level and integrates these approaches to form the client's viewpoint for the delivery of several projects and to enable the sharing of project information across a programme or portfolio of projects for the mutual benefit of all.

References

Antoniadis, D.N., Edum-Fotwe, F.T. and Thorpe, A., (2011) 'Socio-organo complexity and project performance', *International Journal of Project Management*, Vol. 29, pp. 808–816.

Armitt, Sir John, (2012) 'London 2012: A global showcase for UK plc', Department for Culture, Media and Sport, p. 6.

Artto, K.A. and Dietrich, P.H., (2004) 'Strategic business management through multiple projects', in *The Wiley Guide to Managing Projects*, eds. Morris, P.W.G. and Pinto, J.K., John Wiley & Sons, New Jersey.

Baccarini, D., (1996) 'Concept of project complexity – a review', *International Journal of Project Management*, Vol. 14, pp. 201–204.

Barro, R.J., (2009) 'Voodoo multipliers', *Economists' Voice*, February. www.bepress.com/evhttp://www.economics.harvard.edu/app/webroot/files/faculty/7_09_02_VoodooMultipliers_EconomistsVoice.pdf.

Boyd, D. and Chinyio, E., (2006) *Understanding the Construction Client*, Oxford, Blackwell.

CIOB, (2009) *Code of Practice for Project Management*, 4th edn, Ascot, CIOB.

Cox, A., (2004) *Win–win? The Paradox of Value and Interests in Business Relationships*, Stratford-upon-Avon, Earlsgate Press.

Cox, A. and Ireland, P., (2002) 'Managing construction supply chains: A common sense approach', *Engineering, Construction and Architectural Management*, Vol. 9(5/6), pp. 409–418.

Cox, A., (2009) Strategic Management of Construction Procurement', in *Construction Supply Chain Management Handbook*, eds. O'Brien, W.J., Formoso, C.T., Vrijhoef, R. and London, K., CRC Press, London, Taylor and Francis Group.

Dainty, A.R.J., Briscoe, G.H. and Millett, S.J., (2001) 'Subcontractor perspectives on supply chain alliances', *Construction Management and Economics*, Vol. 19(8), pp. 841–848.

Danilovic, M. and Borjesson, H., (2001) *Managing the Multi-project Environment*, The Third Dependence Structure Matrix International Workshop Proceedings, Massachusetts Institute of Technology, Boston, Oct. 29–30.

Egan, J., (1998) *Rethinking Construction: The report of the Construction Task Force*, London, Department of the Environment, Transport and the Regions, currently Department of Trade and Industry.

Engwall, M. and Jerbrant, A., (2003) 'The resource allocation syndrome: The prime challenge of multi-project management?' *International Journal of Project Management*, Vol. 21, pp. 403–409.

Hagan, G., Bower, D. and Smith, N., (2011) 'Managing complex projects in multi-project environments', in Proceedings of the 27th Annual ARCOM Conference, 5–7 September 2011, eds. Egbu, C. and Lou, E.C.W., Bristol, Association of Researchers in Construction Management.

Hossain, L. and Ruwanpura, J., (2008) 'Optimization of multi-project environment', in Proceedings of the 40th Conference on Winter Simulation, eds. Mason, S.J., Hill, R.R., Mönch, L., Rose, O., Jefferson, T. and Fowler, J.W., Miami, The Institute of Electrical and Electronics Engineers. http://www.informs-sim.org/wsc08papers/303.pdf (accessed 14.6.12).

Kim, S.S., Park, J.Y. and Lee, J., (2010) 'Predicted economic impact analysis of a mega-convention using multiplier effects', *Journal of Convention & Event Tourism*, Vol. 11(1), pp 42–61.

King A.P. and Pitt, M.C., (2009) 'Supply Chain Management: A Main Contractor's Perspective', in *Construction Supply Chain Management*, ed. Pryke, S., Oxford, Wiley-Blackwell.

Kumaraswamy, M. and Rahman, M., (2006) 'Applying teamworking models to projects', in *The Management of Complex Projects: A relationship approach*, eds. Pryke, S. and Smyth, H., Oxford, Blackwell.

LEK Consulting, (2012) *Construction in the UK Economy: The Benefits of Investment*, London, Contractors' Group. http://www.ukcg.org.uk/fileadmin/documents/UKCG/LEK/LEK_May_2012_final.pdf.

Loosemore, M., (2006) 'Managing project risk', in *The Management of Complex Projects: A relationship approach*, eds. Pryke, S. and Smyth, H., Oxford, Blackwell.

McCarthy I. and Anagnostou, A., (2004) 'The impact of outsourcing on the transaction costs and boundaries of manufacturing', *International Journal of Production Economics*, Vol. 88(1), pp. 61–71.

National Audit Office, (2001) *Modernising Construction*, HC 87 Session 2000–01, London, The Stationery Office. http://www.nao.org.uk/publications/0001/modernising_construction.aspx?alreadysearchfor=yes.

National Audit Office, (2003) *PFI: Construction Performance*, HC 371 Session 2002–03, London, The Stationery Office. http://www.nao.org.uk/news/0203/0203371.aspx.

National Audit Office, (2005) *Improving Public Services through Better Construction*, HC 364-1 Session 2004–05, London, The Stationery Office. http://www.ccinw.com/uploads/documents/procurement_key_documents/nao_improving_public_services_through_better_construction.pdf.

ONS, (2011) *Construction Statistics Annual, Table 2.1: Value of construction output*, Newport, Office for National Statistics.

Payne, J.H., (1995) 'Management of multiple simultaneous projects: A state-of-the-art review', *International Journal of Project Management*, Vol. 13, pp. 163–168.

Pellegrinelli, S., (2011) 'What's in a name: Project or programme?' *International Journal of Project Management*, Vol. 29, pp. 232–240.

Remington, K. and Pollack, J., (2007) *Tools for Complex Projects*, Hampshire, Gower Publishing.

Rimmer, B., (2009) 'Slough Estates in the 1990s – Client-driven SCM in construction supply chain management concepts and case studies', in *Construction Supply Chain Management*, ed. Pryke, S., Oxford, Wiley-Blackwell.

Sacks, R., (2004) 'Towards a lean understanding of resource allocation in a multi-project subcontracting environment', in 12th Annual Meeting of the International Group for Lean Construction, Helsingor.

Wolstenholme, A., (2009) *Never Waste a Good Crisis: A Review of Progress since Rethinking Construction and Thoughts for the Future*, London, Constructing Excellence http://www.constructingexcellence.org.uk/pdf/Wolstenholme_Report_Oct_2009.pdf.

Williamson, O., (1975) *Markets and Hierarchies*, New York, Free Press.

3

The client's values and the balanced scorecard

View of the Olympic Stadium from under the north end of the Aquatics Centre (photo courtesy of AECOM).

Programme Procurement in Construction: Learning from London 2012, First Edition.
John M. Mead and Stephen Gruneberg.
© 2013 John Wiley & Sons, Ltd. Published 2013 by John Wiley & Sons, Ltd.

Some 4000 trees, 300,000 wetland plants, 150,000 perennials and 60,000 bulbs gave the 2.5 km^2 Olympic Park a lush and green feel for the Games (photo courtesy of AECOM).

Introduction

This chapter describes the starting point for developing a PSE solution to procurement. Defining what represents value is something clients often find very difficult to describe. The chapter therefore discusses the steps required to understand what the programme is seeking to achieve and what needs to be measured in order to demonstrate performance. It equips the delivery team with the means for communicating the value-for-money criteria the client wishes to be met. The chapter also sets out the means to measure value at both the procurement and delivery stages, so that the client can select the most value-aligned contractor for each contract and then measure the outputs of delivery against the same value criteria.

By their nature construction projects are complex, requiring the organisation of many disparate companies and their resources, with numerous specific technical skill-sets. These organisations come together in an ever-changing environment with constantly shifting demands. Construction projects deal with multi-faceted issues, often arising outside the control of their own management. This is a real problem, as many different factors need to be taken into account simultaneously during

A patriotic British gent ponders the scale and transformation of this once rundown part of east London (photo courtesy of AECOM).

project delivery. One of the major challenges for the management of construction work is therefore communicating complicated ideas to the large number of stakeholders that form the chain of command.

Because there is a need to communicate the specific priorities for delivering a construction project and how these priorities fit within the overall strategic aims of a client, it is not sufficient to list sets of requirements, as if they were a shopping list. While clients' own business objectives or reasons for investing capital into the built environment may differ, they require their capital investment to return a construction project that is on time, on budget and of the required quality.

Client organisations are complex and their use of capital assets is diverse. Therefore their wider strategic objectives for making the investment and the requirements expected from the inputs of the built environment can vary widely. Some objectives, such as budgetary constraints, feature more heavily than others for some clients, while other objectives, such as social or employment impacts, may not feature at all. Prioritising these requirements and communicating their relative importance to the supply chain are the starting point of the procurement process. Clients must clearly identify and communicate their expectations, if these are to be met during delivery.

Understanding a client's values and meeting or exceeding them is one of the key criteria for determining success in a project, for both the client and supply chain alike. Ensuring the right questions are asked of the supply chain during procurement enables the selection of the best response that meets these critical success factors.

Some objectives may contradict others, while other objectives may need to take a number of factors into account. Still others may be dependent on circumstances at the time, or depend on progress or decisions yet to be taken. The key issue in communication is to convey a complicated message clearly, consistently and reliably. The lack of completely clear communication is often the root cause of misunderstandings on the parts of both the client and their chosen delivery team.

A key set of issues that need to be communicated at the outset are the client's value criteria, against which they select the supply chain and measure success. These values are made up of certain priority themes and critical success factors, including time, cost and quality. Not understanding these and the client's emphasis on their relative importance can lead to unexpected outcomes, disputes and disagreements. It is therefore necessary to define value for the client in economic, social and environmental terms.

Developing a framework for measuring performance

In order to understand a client's expectations for their construction programme, it is important to assess their strategic goals, their business aims, company mission, vision, values, objectives and targets for improvement. These goals and targets need to be translated into the business case for delivering the built assets of their programme, even though the client may not be in a position to prioritise their objectives or set out their strategic thinking. However, through discussions with the client body, starting at the highest level of requirement, it is possible to establish priorities with increasing levels of detail and eventually weight the key value criteria. By weighting the relative importance of the different aims, the supply chain can respond with their best offer to meet those specific requirements and therefore deliver success as defined by the client's values. The ultimate aim is to develop a framework for measuring performance during delivery and a 'balanced scorecard' that can act as a framework for procurement, against which tenders can be assessed and delivery outputs can be measured.

All from a project vision

The starting point for organising and managing a construction programme, even of the size of London 2012, is to agree a vision statement, which is a broad statement of the purpose to be served. Only when the destination is agreed is it possible to determine the most efficient route to the ultimate objective. Specific themes or 'perspectives' then emerge from a relatively short vision statement.

According to Armitt (2012), there were six priority themes underpinning the construction of the venues and infrastructure. These themes were design and accessibility; employment and skills; equality and inclusion; health, safety and security; sustainability; and legacy.

The programme vision becomes the key statement, against which the benefits of the programme can be measured and it forms the basis for the programme strategy, including the procurement strategy and the management of the supply chain.

The vision statement used for the infrastructure works for the London 2012 games stated:

'To deliver the Olympic Park and all venues on time, within agreed budget and to specification, minimising the call on public funds and providing for a sustainable legacy.'

Once agreed, the vision can be broken down into a number of key perspectives. Breaking down the key perspectives can be carried out in a workshop with the client and key programme management staff, to

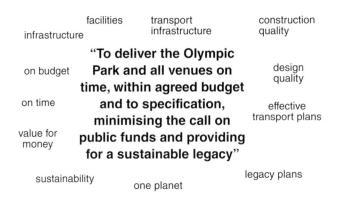

Figure 3.1   The vision and perspectives of the infrastructure works for London 2012.

ensure that there is a common understanding of what the broad vision actually means in terms of more tangible themes or deliverables. Figure 3.1 shows the vision statement and a number of themes that were seen as key issues and that needed to be managed throughout the programme.

Having carried out an analysis of the vision statement, the next step was to establish a number of themes, against which progress toward the vision could be meaningfully measured.

Figure 3.2 shows elements of the vision grouped together to create a number of key themes or perspectives. For example, 'Accountability' refers to schedule and budget, while 'Delivery' refers to the various venues and design quality. Stakeholders and sustainability were also seen as key themes. They proved important, as both these subjects had a major impact on the procurement and supply chain activities of London 2012.

Figure 3.2 also shows that the themes are linked. For example, if labour is developed through training and awareness of sustainability issues, this also contributes towards the sustainable legacy envisaged by the client. If the sustainable legacy is addressed, then the delivery may be seen as excellent, which is of benefit to a number of stakeholders, including the supply chain. This step of the process is one of the most critical, as it begins to identify tangible, linked themes that underpin the client's vision.

Once the key themes or perspectives are identified and agreed, the programme management office team within the delivery partner are in a position to look at the strategic goals within these themes. Again this

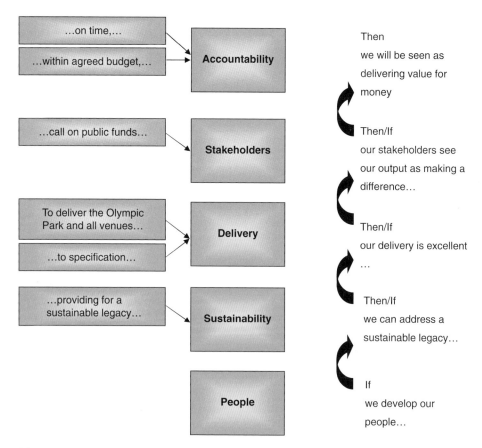

Figure 3.2 The logic and implications of themes derived from the vision statement.

can be carried out in a workshop scenario, which also helps to consolidate a common understanding of the programme goals and how they relate to the client's aspirations and vision.

Each theme or perspective should have between one and three strategic goals. These goals need to be tangible, measurable and understood by the programme team. Also, the goals need to be looked at not only in isolation, but how and why they are linked. Figure 3.3 illustrates the linkages between the various strategic goals within the framework of the themes shown on the left. These linkages are important, as managing performance requires all the measures to be aligned to the ultimate vision.

Once the strategic goals have been agreed and linked, the next step is to assess what success might look like. Critical success factors (CSF)

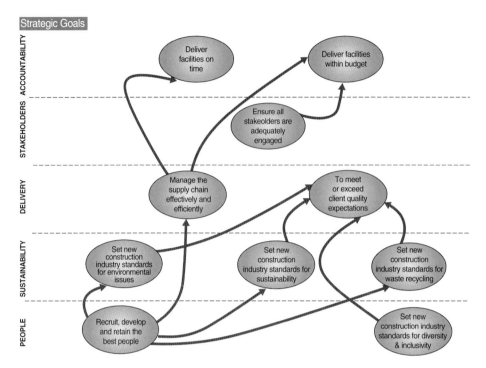

Figure 3.3   The linkages of the strategic goals developed from the vision statement.

need to be identified, as it is those factors that form the basis of the actual measures within the performance management regime. Again, that can be achieved through a workshop environment, where each goal is assessed to identify and understand the relevant and appropriate CSFs. Figure 3.4 shows how these factors can be overlaid onto the strategic goals for presentation purposes. CSFs may include providing clear direction and focus, cost-effective delivery of facilities, facilities meeting requirements and improving customer satisfaction.

Performance measurement

A framework for measuring and managing performance depends on defined criteria for success. Projects and programmes often establish specific and bespoke project measures that reflect one unique project. Unfortunately, these unique measures cannot be used to benchmark or compare with other available data from unrelated projects to understand

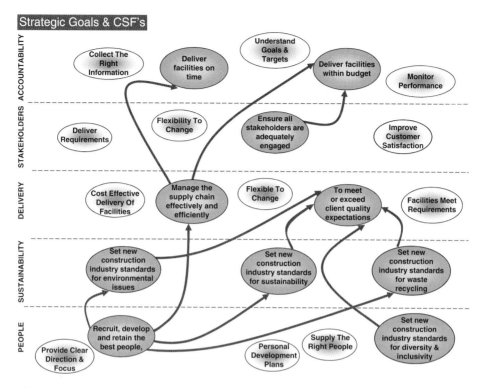

Figure 3.4 Critical success factors and strategic goals.

their relative performance. It is for this reason that the capture and communication of clients' values, their priority themes for delivery and the underlying critical success factors are important.

By defining at the outset what the client requires from the supply chain, it is possible to test the capability of suppliers to deliver during the procurement process. The strengths and weaknesses of their responses to key questions, designed to interrogate their approach to delivery, establishes the benchmark against which they can be measured during delivery.

Tender documents submitted by bidders usually describe technical delivery in great detail and are designed to encourage confidence in the tenderer's ability to deliver. They invariably imply that the tendering firms can deliver technical excellence, assuming sufficient time and enough money. Even though construction supply chains do not always deliver the non-technical priorities of their clients, during procurement bidders offer well-thought-out and detailed responses to these requirements and often score well, giving rise to the hope that they will be forthcoming. Even though what a bidder offers may not actually be delivered, measuring the outputs of delivery against the offers of the tender is often missed as an opportunity to test the benchmark that has been set by the bidding contractors themselves. If an undertaking has been made to do something in a certain way in a bidder's tendered offer, then the project or programme manager should be entitled to expect the supply chain to be able to deliver it. After all, it was on this basis that the supply chain won the contract over their competitors.

Unfortunately, one of the biggest missed opportunities with procurement is that after much time, effort and often substantial resources have been spent on compiling what the tenderer believes to be a model answer that sets out their delivery solution, the tender documents are filed and not referred to during delivery.

In any case, the tenderer's proposal of how local employment might be delivered, or how ethical sourcing requirements of their client, for example, might be achieved, may not be clearly stated and capable of being used as a benchmark against which to measure performance in delivery. If these areas had been probed and tested during procurement and found to be well designed, reflecting the values, priority themes and critical success factors of the client, then as the winning tender they should represent the best response to meeting the client's requirements and therefore they should be used as a measure of success of the tenderer's performance.

The problem with this approach in practice is comparing delivery against tender documents. Making such comparisons can be easily accomplished with tendered cost and time measures, but not quite so easily assessed for many of the more subjective areas. However, using electronic tendering and electronic evaluation to review tender submissions, subject matter experts can be employed to assess specific areas relating to their expertise. Such approaches to evaluation are used in PSE. The advantage is that during procurement the subjective evaluation by the subject matter experts establishes a rating and a rationale that sets out why a score has been awarded.

If these specialist evaluators are drawn from the same team that then manages the delivery of their particular area of expertise, then regular reviews of tender documents compared with the contract during delivery can establish whether the actual delivery does indeed reflect the solution offered in the tender documentation. Electronic systems make regular assessment during delivery feasible and help to review performance trends across a programme of projects, capturing best practice.

Sharing best practice across a programme enhances delivery across teams in a shared learning environment, to the mutual benefit of client and supply chain alike.

Although benchmarking and sharing good practice across a programme may be seen as a threat by some, the argument in favour is that a programme of ten projects would not be seen as successful, if only nine out of ten of the projects met the requirements of the client. Performance improvement is driven by competition across the supply chain for a programme in terms of transparency of performance, peer pressure and the promise of rewards for good performance. Measuring performance using the same criteria of a client's critical success factors across a number of teams and timeframes allows comparisons of projects to be made and assists in achieving overall programme success.

Using balanced scorecards to communicate values and measure performance

A balanced scorecard is used to communicate the values of a client to the supply chain. The scorecard sets out a large number of aims and objectives on a grid or matrix. The ODA developed the scorecard shown in Figure 3.5, which highlights the priority themes and a number of critical success factors for delivering the objectives and ultimately the ODA's stated mission. Each of these objectives and critical success factors is underpinned by a standard set of key performance indicators (KPIs), which are measured and used to manage performance during delivery.

The aims are prioritised or balanced in terms of their relative importance by weighting each objective. The balanced scorecard is divided into detailed areas for testing during procurement, with points awarded for each question, weighted according to its relative importance to the client. Each contractor is assessed or rated against the scorecard and can then be compared with all the other bidders to identify the strengths and weaknesses of each.

Several key themes are shown in the balanced scorecard in Figure 3.5, including costs, which need to be sufficient to facilitate cooperation within the supply chain and provide enough incentive to deliver client values. Costs should be managed collaboratively and the specific requirements defined accordingly. A second theme involves timing or scheduling to ensure that supply and delivery are aligned to a master schedule,

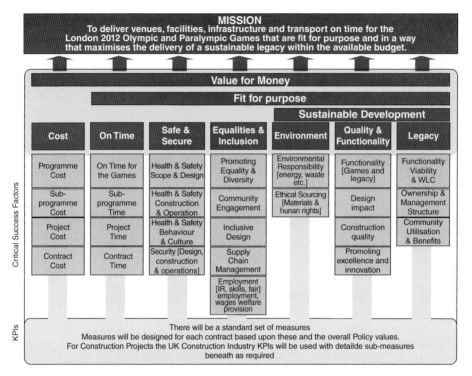

Figure 3.5   ODA Balanced Scorecard.
Source: Olympic Delivery Authority – Learning Legacy Balanced Scorecard – Champion Products (October 2011)

so as to deliver the overall objectives and benefits of the programme within the timescale envisaged and agreed.

Quality, as a theme on the Balanced Scorecard, is one of several critical success factors used to define quality and possibly includes the use of the industry-developed design quality indicators (DQIs), which might be used to measure and demonstrate design excellence. Several themes come under the broad heading of sustainable development, including environmental sustainability to ensure that natural resources are used as efficiently as possible, while setting particular environmental targets that are appropriate to the programme. Similarly, social sustainability is included as a theme to ensure that both the programme and the supply chain contribute to the social improvement or regeneration of particular areas and perhaps leave a long-lasting legacy following the conclusion of the Olympic Games. The purpose of economic sustainability is to ensure that the programme contributes to creating a sustainable market not only by supplying goods and services to the programme, but also by creating employment and training opportunities and by assuring minimum wage and other general working conditions are applied on all projects.

Balanced scorecards can be used to communicate many of the priorities of clients. These priorities translate into policy objectives that guide the tender process and subsequent contracts. Balanced scorecards enable delivery to be managed and measured across different policy areas. Scorecards can also be applied further down the supply chain. For example, tier 1 contracts with main contractors may include a requirement to apply some or all of the same balanced scorecard to all tier 2 procurements with subcontractors. This ensures that the client's policy objectives are cascaded down the supply chain to smaller firms, who might not otherwise have been directly exposed to client-led initiatives. In this way scorecards can help to monitor, manage and control tier 1 and tier 2 contractors during procurement and delivery.

This scorecard method can be used to raise issues and aid communication on a variety of matters, including cost certainty and life-cycle costs, to establish which matter most to the client. For example, a grocery retailer procuring a building might be most interested in their built asset making the fastest return on investment, while a care-home provider might be looking for longer-term savings on the running costs of the facility and the initial capital cost per room.

One of the major client objectives is very often timely delivery. The most common remedy for failure to meet this requirement is the use of liquidated and ascertained damages (LADs). Although Lal (2009) points out that the advantages of LADs are that they are precise and unambiguous, others, including Bingham (2008), have argued that they are seen by contractors as penalties waiting to happen and can, therefore, cause more problems and disputes than they solve.

This implies a need for clarity and a need to communicate in precise terms, as early as the initial procurement stages, a vision of what the client demands. The London 2012 programme involved a major procurement exercise spanning a number of years with a large number and variety of stakeholders, whose needs, expectations and aspirations needed to find a balance of requirements using a scorecard. These requirements included security, employment, environment and legacy objectives.

One of the features of very large construction projects is that they have a major and transforming effect on the thousands or even millions of people who are impacted by them. In this sense, procuring projects on the scale of the London 2012 Games has impacts on their environment, employment and quality of life. In an important sense, major projects give decision makers responsibility for far more than simply acquiring buildings and structures. It is no longer simply a matter of just buying buildings. Major projects create a way of life and leave a legacy; London 2012 recognised this and made that aspect a part of their strategic thinking.

Developing a balanced scorecard

In a simple set of objectives consisting only of cost, time and quality, an emphasis on any one of the three would detract from the other two. In other words costs could be reduced at the expense of project duration or by compromising quality. However, because of the addition of a number of complex objectives caused by environmental impact studies, health and safety legislation and global warming, a matrix of competing objectives needs to be delivered. These objectives have to be budgeted for during the business case, clearly articulated during supplier engagement, weighted and evaluated during procurement and then measured or benchmarked during delivery. To manage all of these processes requires a coherent contracting strategy that clearly sets out the responsibilities, when they arise and where they are located. Without this direction the supply chain cannot know what represents value to the client and therefore how best to frame their tenders to deliver best value.

It is therefore necessary to communicate items beyond the building specifications to meet the client's requirements. Defining what is needed sets the agenda for the programme or project. It is this value definition that is the starting point for developing a coherent and clear procurement and supply chain management strategy and it is this definition of value that PSE uses to engage with the client and key stakeholders.

Clients may have some very clear ideas and definitions of what represents value or what a successful programme may look like from their own perspective. They may state the need for their projects to be delivered on time, to budget and to a high-quality specification. Safety issues have become more prominent as a result of legislation. Other objectives

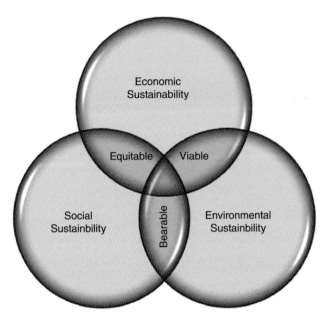

Figure 3.6   Triple-bottom-line sustainability model.

or criteria of success may also be added – for example, there may be issues that need to be delivered by the capital investment, including training and employment, responsible sourcing of materials, engagement with local communities and the creation of a legacy through long-term employment opportunities.

When framing these multi-value criteria, the triple-bottom-line sustainability model, described by Elkington (1994), is useful to bear in mind. The triple-bottom-line model, shown in Figure 3.6, refers to the economic, social and environmental implications of projects. It encompasses several different values that a client may require and helps to translate these success criteria into tests that can be applied during the whole procurement and delivery cycle. PSE uses the triple-bottom-line model of sustainability as the starting point for understanding a client's requirements at the highest level.

The economic, social and environmental aspects of construction projects are often key measures of success. In the UK, Constructing Excellence (CE), an industry body, has framed its KPIs around this model and has been benchmarking and measuring demonstration projects' performance in these categories since 2000. While the CE KPI measures may be very high-level and not ideal for all scenarios, they do help to establish targets and best-practice benchmarks.

The priority themes from a public-sector client's perspective need to be consistent with the key policies that contribute to the political vision of the investment and, similarly, from a private-sector client's perspective the themes need to be consistent with company public-relations values. These priority themes are often the first stage in identifying and prioritising the value criteria that underpin the vision and values of a construction programme. Under each of these priority themes, more detailed critical success factors can be devised to meet the aims of a client's programme.

Measures including key performance indicators

A scorecard of a number of key objectives and underlying values is used to monitor performance across the 'triple bottom line'. It is essential that these values are clearly communicated during the procurement process and that the selection of the supply chain is based on its ability to deliver them. The alignment of the values and objectives of both client and supply chain facilitates the achievement of the stated values.

At a strategic level a major global event or programme such as London 2012 can be benchmarked against other programmes delivered by the construction industry, utilising the available Constructing Excellence industry KPIs. These KPIs cover the three aspects of the triple bottom line.

Construction KPI measures of economic sustainability

- Client satisfaction – product
- Client satisfaction – service
- Defects
- Predictability – cost
- Predictability – time
- Construction time
- Construction cost
- Safety
- Productivity
- Profitability.

Social sustainability: Respect for people KPI measures

- Employee satisfaction
- Staff turnover
- Sickness absence
- Safety
- Investor in people
- Working hours
- Pay
- Training
- Equality and diversity
- Qualifications and skills.

Environmental sustainability: Environmental KPI measures

- Impact on environment
- Energy use (designed) – product
- Energy use – construction process
- Mains water use (designed) – product
- Mains water use – construction process
- Whole-life performance – product
- Area of habitat created/retained – product
- Impact on biodiversity – product
- Impact on biodiversity – construction process
- Commercial vehicle movements – construction process
- Waste – construction process.

Each of the CE KPI measures listed above includes a data-collection method and a defined method for measuring the indicator. In the London 2012 programme each of the KPIs was allocated an owner for that specific area of performance. The data was then centrally coordinated by the programme management office (PMO), one of whose functions is to organise reporting processes. The KPIs were then passed on to the individual teams across the whole programme management spectrum at both programme and project level, to be used by them to manage performance. By comparing individual KPI measures against industry performance and across supply chains both vertically and horizontally in the programme, performance was managed strategically. In this way detailed operational KPIs can be used to pinpoint specific failings requiring remedy.

The balanced scorecard includes a balance of objectives across the 'triple bottom line' agenda, and the operational KPIs are then used to measure performance against the strategic aims. There are a number of detailed operational measures, including the KPIs, and each is comprised of the following components:

- measure – what is being measured (e.g. the ratio of delivery to cost)
- definition – a short description of what the measure captures (e.g. a comparison of budget and outturn costs)
- metric – the unit of measurement to be used (e.g. the percentage variance)
- objectives – the reason why the measure is being used (e.g. to monitor the costs and ensure delivery to within budget)
- target – defines target to be achieved (e.g. one per cent below budget)
- owner – who is responsible for collecting, monitoring and reporting the measure, (e.g. the cost controls team or site supervisory team)
- source material – identifies from where the data is to be collected (e.g. project cost reports)
- frequency of measure – how often data is collected and measured (e.g. monthly)
- frequency for reporting – how often the measure is reported to relevant stakeholders (e.g. monthly)
- enablers – what steps or actions are required to enable measurement (e.g. agree budget, allocate responsible staff, scope of measure and report)
- success criteria – define the minimum satisfactory threshold and above and the impact on the programme.

Together these elements ensure that the measures serve a practical purpose. Using those elements and taking budget compliance as an example, Table 3.1 shows how an operational measure is constructed.

The use of predefined KPI measures enables the programme management team to monitor a number of areas, including performance, the use of incentives and the effectiveness of supply chains. Performance can be monitored and corrective action plans put in place before poor performance impacts negatively on the wider programme. The KPI measures can be used to develop incentive schemes, which both reward good performance and punish weak output using a gain/pain mechanism. The effectiveness of a supply chain can also be compared with others. By using the same measures across a number of supply chains,

Table 3.1 Example of an Operational Measurement

Measure	Definition	Metric	Objectives	Target	Owner	Source Material	Frequency Measure	Frequency Reporting	Enablers	Success Criteria
COST Budget Compliance	Composition of forecast outturn cost (FOTC) against agreed budget	Index based on FOTC divided by budget	To ensure the outturn cost is managed within budget	<=1	Controls	Agreed budget and cost reporting system	Monthly	Monthly	1. Agree budget 2. Agree earned value analysis method 3. Establish cost management system	Pragmatic system to accurately monitor costs against the budget

the suppliers can be benchmarked. KPIs can capture best practice demonstrated by good-performing suppliers. Good practice can then be used to inform and coach poor performers to promote collaborative working, supplier development and overall performance improvement.

Creating appropriate KPIs from a project vision and scorecard

The objective is to create a suitable number of KPIs to encourage behaviour and performance to meet the client's aspirations for a complex programme of construction. Having established the strategic goals and associated critical success factors (CSFs) at the beginning of the programme, KPIs need to be identified that underpin the strategic goals and relate to the CSFs. These KPI measures are linked to the vision statement shown at the top of Figure 3.7. For example, the vision statement includes delivering a sustainable legacy. That translates into a number

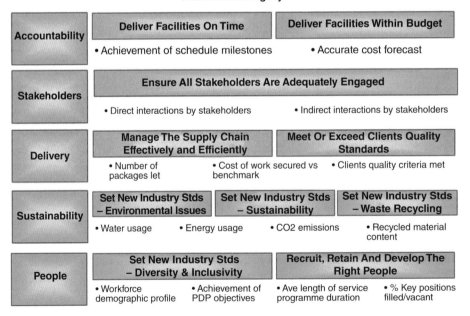

Figure 3.7 Performance indicators used on the London 2012 Infrastructure programme.

of measures, including for example water usage, CO_2 emissions, recycled material content and workforce demographics.

Performance scorecards can be derived from the list of measures, as shown in Figure 3.7. On the left side of the diagram are the key themes or perspectives, ranging from Accountability to People. Each of these key perspectives has between one and three strategic goals. Beneath each strategic goal are the measures used to assess performance on a scorecard.

Concluding remarks

Developing an understanding of a client's requirements is fundamental to delivering a successful programme or project. This is the starting point of the procurement of a construction supply chain. PSE therefore begins by assessing these requirements and then quantifies their relative importance using a balanced scorecard. The balanced scorecard sets out specific areas of performance that can be thoroughly tested during procurement to enable the selection of the best response from contractors and suppliers. Their performance is then tested throughout delivery to ensure that delivery is consistent with what was promised during the tender process.

Procurement may be seen, by firms tendering for work, as costly and often repetitive. It is also seen as adding cost to delivery. Even unsuccessful bidders have to recoup their costs of tendering by imposing the hidden cost of failed tender bids on other clients engaging the failed bidder. However, public-sector procurement is required to be transparent, fair and equitable. Even private-sector clients need to be reassured about what they can expect to achieve with their built environment investment. The failure of a project or programme may be caused by poor procurement, through focusing on misleading criteria, such as selection driven by a lowest price, or not clearly establishing in the first place what was expected of the delivery team.

An unambiguous set of requirements that sets out the relative value of each aspect of a programme allows the supply chain to make a commercial decision on what price they decide to allocate to delivering each requirement. The description of requirements also enables responding contractors and suppliers to state their interpretation of that delivery and its price. Measuring their delivery against their stated intent can be used to enhance performance, engender innovation and achieve continuous improvement across a programme.

Devising a framework for measuring performance – the Balanced Scorecard	
Dos	**Don'ts**
Engage ALL stakeholders and establish their objectives	Dive into detail on specific contracts
Present the aims and objectives to the stakeholders during engagement	Be steered by the loudest voice
Capture and review all policy/business case documentation	Prejudice the answer
Establish constraints	Expect to please everybody
Agree a timescale and set a timetable	Overcomplicate the answer/outputs
Identify and establish who is responsible, accountable, consulted and informed (RACI)	
Carry out a gap analysis	
Keep it simple	

References

Armitt, Sir John, (2012) 'London 2012: A global showcase for UK plc', Department for Culture, Media and Sport, p. 6.

Bingham, A., (2008) 'It's a lads' thing', *Building*, 28 November.

Elkington, J., (1994) 'Towards the sustainable corporation: Win–win–win business strategies for sustainable development', *California Management Review*, Vol. 36(2), pp. 90–100.

Lal, H., (2009) 'Liquidated damages', *Construction Law Journal*, Vol. 25(8), pp. 569–590.

4

Packaging and contracting strategies

Velodrome structural steelwork during construction, sitting on more than 900 pile foundations (photo courtesy of Mark Lythaby).

Programme Procurement in Construction: Learning from London 2012, First Edition.
John M. Mead and Stephen Gruneberg.
© 2013 John Wiley & Sons, Ltd. Published 2013 by John Wiley & Sons, Ltd.

The completion of 5000 m² of Western Red Cedar cladding brings the Velodrome's stunning curves to life (photo courtesy of Mark Lythaby).

Introduction

Having discussed the aims and priorities of the client in a 'balanced scorecard', described in the previous chapter, we turn to the processes involved in enabling purchasing activities to be undertaken with confidence, bearing in mind the management of risk. The chapter begins with a discussion on packaging strategy. Gestalt theory in psychology is then used to explain why a clustering approach is an effective method of comprehending many projects without necessarily abandoning the characteristics that make each project distinct.

Sixteen kilometres of cable being readied to construct the Velodrome roof, one of the UK's largest examples of a cable-net roof (photo courtesy of Mark Lythaby).

In this chapter we explore how the purchasing of complex projects in a large programme can take advantage of various economies of scale by clustering, which allows for the benefits of repetition and ease of management at programme level. On the London 2012 programme six different clusters were considered, and these then formed the basis of the work breakdown structure and packaging strategy.

The following section uses these clusters and work breakdown structures to devise the programme schedule linking all projects within the programme. With the programme packaging agreed, the next step is to develop a contracting strategy. This varies, as different types of project entail differing levels of perceived risk, which affect the appetite of firms to become involved in the tendering process. The contracts therefore have to be sufficiently attractive to encourage firms to bid. On the London 2012 programme, the New Engineering Contract 3 was adopted in the main, with several options for the different types of project or cluster to be undertaken. These options are described in detail.

What, why and how to buy

Procurement can be inefficient and burdensome for organisations that are required to establish procedures that are transparent, auditable and compliant with legislation and the rules and regulations set by the requirements of the client. The concept of Purchase and Supplier Engineering (PSE) is distinct from the term 'procurement', because it is more than simple procurement. The additional elements described here can improve the efficiency of the procurement process.

Construction procurement for one-off projects is often bespoke and developed either in full or in part for each project. In contrast, the purchasing function of PSE is process-driven and designed to be efficient through standardisation and repetition, much like a production line. PSE offers a standardised mechanism to buy construction efficiently. It recognises that certain elements of the procurement process can be standardised, while allowing other specific elements to be tailored to meet particular one-off requirements.

The PSE model was developed and first used in the procurement of the construction infrastructure for the London 2012 Olympic and Paralympic Games. While it was not known as PSE at the time, it was developed by the delivery partner and their specialist team of procurement and supply chain managers.

After deciding needs and requirements the starting point for any purchaser is what to buy. This is true for any shopper, from household shopping to procuring a large construction programme. The actual buying processes may be very different and construction projects may be very complicated, but the purchasing questions remain the same: what to buy, how best to buy it and where to buy it. For a grocery shopper it may start with compiling a shopping list, followed by selecting which supermarket or shops need to be visited. Physical visits to shops are, of course, increasingly being done 'virtually'.

Compiling a shopping list for a construction programme requires far more strategic thinking than a weekly domestic shop. On a construction programme, the PSE process begins with a distinct set of meetings to develop requirements. This phase of developing requirements involves meetings with the various stakeholders and the agreement of the client body for a balanced scorecard of value objectives. The aim of the programme purchaser is to ensure that the programme is bought in such a way as to meet the value objectives agreed during the Develop Requirements stage of PSE. At one extreme this may be a large engineering, procurement and construction (EPC) contract, requiring a major construction organisation or joint venture to take on the risk of engineering and construction. Alternatively, the client could decide to break up a programme into discrete packages or categories of work to manage the risk via separate individual contracts. This approach enables the client to maintain a greater overview and achieve greater control over the delivery of packages, ensuring the client's own due diligence is applied during the procurement of main contractors and then through the contracting strategy to the critical firms in the supply chain. It was this latter approach that was chosen by the Olympic Delivery Authority (ODA) in 2006, when they appointed a delivery partner to manage their construction programme.

However, the challenge remains: how does one buy an Olympics? Where does one start? The Olympics required a wide-ranging cross section of construction skills. The packaging and contracting strategies had to be designed to ensure that the appropriate packages were contracted to the most capable suppliers, while allocating risk to those best placed to manage it.

Devising a suitable procurement strategy depends on a strict, logical order. This is illustrated in Figure 4.1, which illustrates the procurement cycle. The first step is to gain an understanding of what exactly needs to be bought. The requirements for London 2012 were well established. How to buy those requirements, or bundle them up for procurement, formed the packaging solution for the programme's requirements – the second step in the procurement cycle. Only once this was complete was the problem of how best to transfer risk considered – the third step in the cycle. This was achieved through developing a contracting strategy. With the packaging and contracting strategies in place, individual packages or project procurement plans could then be tested to gain an understanding of market appetite, the interest and willingness of firms to compete during procurement and participate in the delivery of the programme.

PSE adopts this strict order in developing a programme procurement strategy, because the packaging strategy takes into account the overall

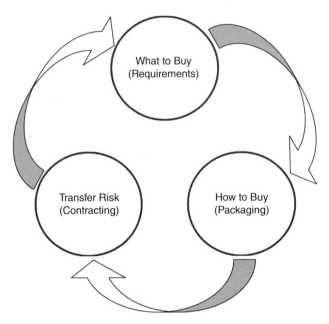

Figure 4.1 The procurement cycle.

requirements, the scope and scale of packages and their associated risks, and the interfaces between packages. The apportionment of these risks can then be matched with the most appropriate contracting strategy. Packaging decisions are about establishing the boundaries of risk, while the contracting strategy transfers and allocates those risks accordingly. The contracting strategy is invariably linked to the need for incentives and the overall delivery objectives.

Packaging strategy

The purpose of a packaging strategy is to plan and coordinate the construction delivery undertaken by different firms to meet the requirements of a programme. The programme sets out the priorities of what exactly is to be delivered, a timetable for each facility and the infrastructure needed.

The packaging strategy is developed by breaking down the requirements of the whole programme into its constituent parts. This allows the overall delivery requirements to be simplified to facilitate the development of specific procurement plans for design, construction works and services.

There are many benefits to be had, if a programme management team adopts an overarching strategic approach early in the process. An early understanding of programme deliverables, attendant risks and constraints helps the programme management team to be consistent in their approach across the programme on all projects and packages.

Knowing what is to be procured and how these procurements are to be managed is essential for gaining the confidence of stakeholders, both on the supply side and on the demand side of the process. Once that confidence is established, the programme can be developed with a sound set of principles for delivery, with the support of both the senior management and the client's executive team.

The larger the package, the less visibility and transparency is available to the client, because the greater the fragmentation in the supply chain tends to be the larger the supply chain. Packaging large construction procurements inevitably requires subcontractor involvement, with implications in terms of their capacity and capability. Over-commitment by tier 2 contractors, and the fact that tier 1 contractors may not be aware of the outside commitments of their own subcontractors across the whole programme, can lead to delivery and company failures. Programme risk exposure to critical tier 2 contractors is a constant, ever-changing issue that needs to be continually monitored. Where capacity or capability is being stretched, warnings need to be given by the programme supply chain managers to the procurement team, so that planning can be revised or adjusted accordingly.

Tier 2 market engagement in the planning and packaging of work needs to take account of the market's ability to deliver the works. This is also indicative of the validity of the supply chains assembled by the tier 1 contractors.

The development of a packaging strategy involves planning how the construction delivery is to be undertaken in relation to the requirements of the programme. It takes into account the scope and scale of each package, the interface between adjacent projects, the delivery risk and the schedule for completion. The programme sets out the priorities of what exactly is to be delivered and a timetable for each delivery item, including the necessary infrastructure to enable and assist in that delivery. This timetable is then used to inform the procurement strategy and drive the procurement plan. The tender event schedule (TES), shown in the upper part of Figure 4.2, is an example of a timetable against which the progress of each package procurement can be monitored.

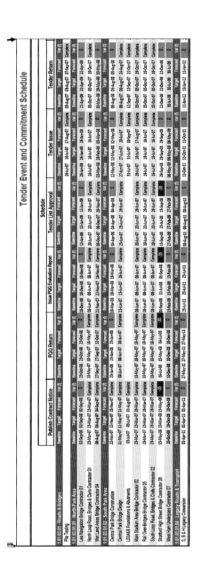

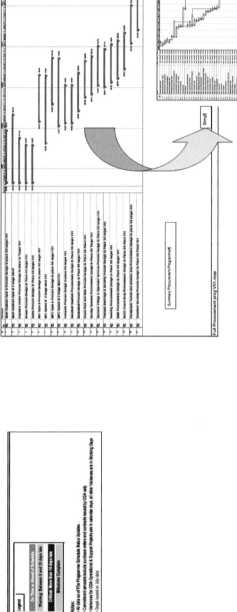

Figure 4.2 Tender event schedule, programme schedule and procurement strings.

There is a requirement, very early in the process, for the core programme procurement team to assist in deciding the outline and the scope of the programme and the construction sequence. The programme schedule must meet the overall completion date. Procurement time periods, called strings, are shown in the lower right of Figure 4.2. These strings are developed for each package as part of the overall scheduling requirement, and allow schedule planners to determine practical start and finish dates for each package. The procurement strings are modelled on the timeframes necessary to conform to the obligations of the Public Contract Regulations and the specific requirements of each package.

Gestalt theory

In the field of psychology optical illusions are often used to demonstrate that patterns of thinking influence perception. For example, Figure 4.3 illustrates two horizontal lines of exactly the same length: yet the upper line appears longer than the lower line, the only difference being that one line is shown with the chevrons pointing outwards and the other with the chevrons pointing inwards.

Gestalt theory suggests that human perception attempts to group separate items together to form patterns. These patterns form shapes and hence the theory is known as gestalt theory, after the German word meaning 'shape'.

According to gestalt theory our brains see patterns, even where objects are independent of each other. We organise what we see into groups based on a number of principles such as proximity, where individuals are close; similarity, where individuals have common features; closure,

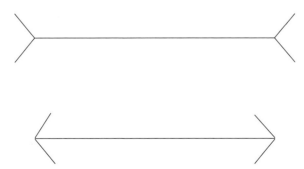

Figure 4.3 An optical illusion.

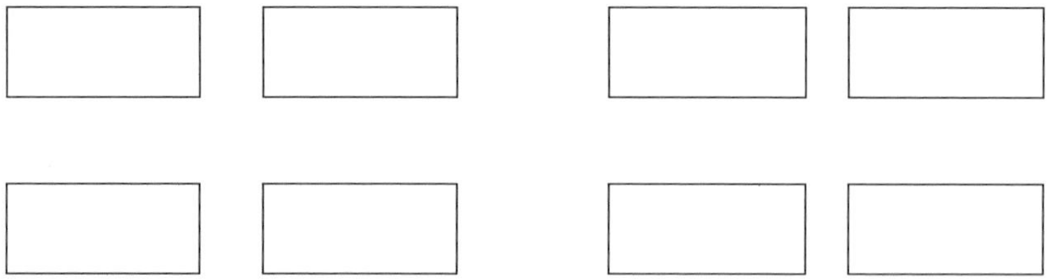

Figure 4.4 Gestalt diagram of eight independent rectangles arranged in two groups of four.

where individuals form a circle, for example, with a gap on the circumference. Because of closure, the gap is ignored by the brain and the circumference is still perceived as a circle. As the theory is outside the scope of this book, a seminal text on gestalt theory can be found in Ellis (1938) or later in Palmer (1999) or in any general introductory textbook on psychology.

The reason for introducing gestalt theory here is that the silo mentality tends to separate the management of projects into isolated units, instead of seeing or looking for patterns of relationships that can be found in the interdependencies of projects within programmes. Take, for example, the eight rectangles in Figure 4.4. The eye immediately interprets the arrangement as two sets of four rectangles. The individual shapes are not seen as eight independent rectangles, but as part of a pattern, each related in some way to the other seven. Each shape is in the same group as another three, but in a different group from the other four rectangles. These are unseen relationships, which help to interpret the pattern in the mind of the observer.

In the same way, each project has a set of relationships with the other projects in a programme, and it is important to define these interdependencies in order to manage them effectively. Moreover, as can be seen from Figure 4.3, it is possible to find that the whole is more meaningful than the sum of the parts, though the interpretation we place on the whole may not necessarily be correct, bearing in mind that both horizontal lines in Figure 4.3 are identical but appear to be different.

In Figure 4.5 eight projects are arranged to show that projects A to D may have connections they do not share with projects E to H. Their relationship may depend on a common source of materials, or it may be that the same tier 3 contractor is engaged on projects A to D but not on E to F. The simple lesson here is that it is not helpful to deal with projects in a programme in isolation, as if they are all in separate silos.

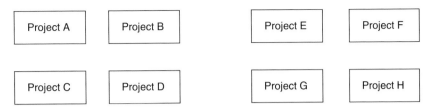

Figure 4.5 Diagram of eight projects arranged in two groups of four.

Of course in reality there are many strands, both contractual and physical, that link projects within a programme, so that it is important to consider the implications of the effect that projects have on other projects in an interlocking programme.

Programme clusters

According to Cornelius *et al.* (2011), a programme clustering model may be used to facilitate the development of procurement strategies for both construction works and design. The use of clustering, combined with a standard set of contract solutions, allows control of the contracting process to remain at programme level and breaks down complex projects into more manageable steps.

On a diverse programme with many contracts involving many firms, there is a need to combine projects to make them capable of being managed and to simplify the overall supervision of the vast array of work in different locations. Otherwise the work of overseeing a large programme becomes far too complex to manage, coordinate and control. Programme complexity can be overcome by clustering projects; for example, clusters were developed to apply to all Olympic capital works projects both on and off the Olympic Park.

Cornelius *et al.* (2011) describe two types of cluster: namely, service clusters and construction clusters. Service clusters, according to Cornelius *et al.*, are used to supply products to all projects, such as street lighting, materials like concrete or timber and even security on all building sites. These provisions are not considered to be clusters in their own right, as they are used only to service the firms undertaking the actual construction work.

The second type, construction clusters, combine construction projects in the same location or close proximity that could benefit from contractors collaborating. The clustering approach is sufficiently flexible that it can accommodate additional projects, as and when the need arises. This is the clustering model that was adopted in the London 2012 programme to simplify delivery requirements, taking advantage of economies of scale where appropriate and benefiting from the cross-fertilisation of ideas and experience. Clustering projects helped the management team to communicate with the supply chains that were charged with delivering within the clusters. The use of clustering also helped to determine a set of standard procurement and contracting routes. The clusters shown in Figure 4.6 were created by taking the Olympic programme and separating the layers sequentially, in the approximate order in which structures were to be built. The characteristic of each cluster of projects was that each represented a form of infrastructure or building. There were six construction clusters in all, consisting of:

1. Enabling works
2. Utilities
3. Structures, bridges and highways
4. Permanent venues
5. Temporary venues, and
6. Landscape and the public realm.

These clusters were then used by the programme team to devise a work breakdown structure (WBS), which is a method commonly used in project management to divide complex problems into manageable portions. That is done by defining the extent and scope of the whole structure to be built before creating a hierarchy of building elements. These building elements are in turn broken down into their parts, and some may require a further breakdown. Methods of construction, complete with a budget, a named responsible manager and target dates for completion, then define the work package, which contributes to the final total building solution. By dividing the whole building or structure into its elements, each work package must ensure that nothing is omitted from the client's stated outcomes. The scope of the total work completed must represent 100 per cent of the client's requirements.

The WBS enables the construction work to be packaged, using a project dictionary to define the scope of each project within the clusters. The WBS can then be used to outline each individual procurement into specific projects and work packages. The work must be packaged before deciding the contracting route, because the work packages dictate the type of contract required for the purpose of allocating and managing risk.

A clustering approach is used for consistency. It helps to determine a standard procurement and contracting route and to assess risk in the contracting strategy. A number of factors need to be considered when clustering packages, including the following:

• the timing of the works
• the location of work in relation to other works, i.e. the interfaces, and
• the construction techniques and the technologies to be used and their supply chains.

What should also be considered is whether the cluster is likely to offer opportunities to share logistics, knowledge, experience, resources and expertise, and also whether the market has the capacity and resources to meet the demand created across the cluster and the programme.

Figure 4.6 shows the order in which project clusters were to be delivered on the London 2012 programme – often referred to as the 'layers of cake', with the landscape and public realm cluster representing the

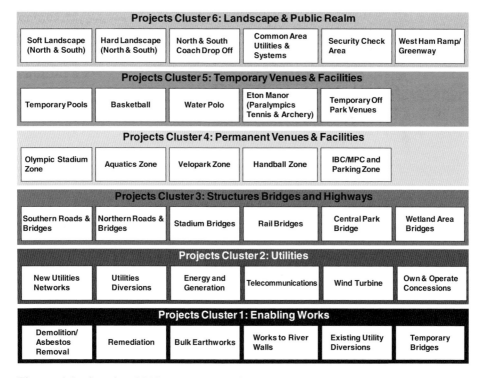

Figure 4.6   London 2012 programme clusters.

icing on the cake. These clusters are strategic categories that define the scope of works and services. They outline the approach to be adopted when defining the discreet packages for procurement and form a cluster strategy. Work has to be packaged before deciding the contracting route to procurement, as the work packages dictate the type of contract required for the purpose of managing risk.

Programme application

Using the work breakdown structure (WBS) and procurement strings described earlier, seven procurement milestones were identified. These were:

1. Procurement plan approved
2. Contract notice and pre-qualification questionnaire (PQQ) published
3. Tender list authorised
4. Invitation to tender (ITT) issued
5. Tenders returned
6. Award recommendation, and
7. Contract awarded.

These procurement strings were then used on the London 2012 programme to link anchor milestone dates and to develop the tender event schedule (TES). The TES was used to monitor procurement progress and compare actual progress with the agreed overall programme forecast. The TES was used extensively by senior managers within the ODA and the delivery partner as a high-level reporting tool and for departmental resource planning. In addition, the data was used to inform and engage the supply chain through CompeteFor (an electronic brokerage website used to promote emerging London 2012 supply chain business opportunities) and the London 2012 website. These websites openly published what was to be procured and those procurements that were already under way, and in the case of the ODA's direct procurements the names of firms that were short-listed and invited to tender.

The publication of approved tender lists at milestone 3 gave the lower tiers of the supply chain the opportunity to engage with the short-listed tier 1 contractors. By providing the names of main contractors, subcontractors were given the opportunity to bring their specific expertise or innovation to the solution. They were therefore able to offer their services directly to tenderers in the development of their submissions. This approach also enabled the team to fulfil its transparency obligations and provide small and medium-sized enterprises (SMEs) with the visibility of potential openings, which they could target to assist their entry into the supply chain.

In order to gain and retain the confidence of stakeholders and the executive body, an early understanding of what is being procured and how the works are planned is essential. It is also important to maintain that confidence by accurately reporting the current progress of key procurement activities.

Contracting strategy

At the beginning of the London 2012 procurement an early analysis of the clustering approach and the WBS indicated the need to procure a large number of construction and related contracts in a very short period of time. To put this into context, the programme showed that the ODA procurement function had to go from almost no spending to a £1 bn buying function within a year. To satisfy complicated contracting requirements a standard approach to contracting had to be adopted before specific projects could be procured. The approach taken was to develop a contracting strategy based on a classification structure, which responded to the associated risk profiles of each project.

The formal procurement policy set out a requirement for robust but flexible contract arrangements that took into account the levels of risk over the duration of the entire programme. The contracting strategy responded to this requirement by using the New Engineering Contract, 3rd edition (NEC3) suite of contracts to formalise arrangements.

Forms of contract used in the 2012 Olympics procurement

Selecting appropriate contract forms is the major component of a contracting strategy. The New Engineering Contract (NEC) was first published in 1993 and in 2005 a revised version, NEC3, was introduced. The NEC3 suite of contracts, comprised of a number of contracts with various options designed to meet a variety of situations and purposes, was the preferred form of contract for construction projects on the

London 2012 Olympic and Paralympic Programme. This was principally because they were seen as collaborative forms promoting construction industry best practice, while also offering a range of flexible solutions for contracting out various levels of risk.

The NEC3 contract encompasses and integrates a set of processes that ensure corporate governance. For example, as a publicly funded organisation it was incumbent on the ODA to manage public funds appropriately and transparently, when initiating contract changes during the contract period. Variations could be managed within the compensation event process, within the NEC3 contract. The early warning notice procedure of the NEC3 also provided a contractual method for both parties to the contract to mitigate risk. NEC3 contracts can be used worldwide and are indeed used in many countries.

Although the NEC has been viewed as onerous to administer, it encourages a collaborative and proactive approach to contract administration and ensures a focus on time as much as on cost. It was shown on the London 2012 programme that NEC contracts could be managed and administered with very few disputes. In general the contracts were well administered, although that owed as much to a change in the organisational culture and approach of all involved as it did to the availability of experienced individuals to manage and administer the programme appropriately.

The London 2012 construction contracts were standardised, using the NEC3 suite of contracts to provide a consistent approach to contracts across the programme and throughout the supply chain. This was achieved by means of a 'cascade' approach to subcontracts, where key clauses were embedded in the tier 2 and 3 subcontracts, including payment terms. The NEC3 consists of a suite of contract options, A to F, with each option to be read in conjunction with core clauses. The options adapt the NEC3 to the works or services required. The preferred options used were:

- Option A: Priced Contract with Activity Schedule
- Option C: Target Contract with Activity Schedule
- Option E: Cost Reimbursable Contract
- Option F: Management Contract
- NEC Term Service Contract.

These options, which are reviewed below, allowed the ODA to select the most appropriate risk profile, while balancing time, cost and quality.

The NEC3 Option A contract is a priced contract with an activity schedule. Where possible the fixed-price Option A form was preferred. This was the primary contracting route for assets that could be well defined and where cost and time risks could be identified and appropriately managed. This option provides value for money and precludes high-risk premiums. Option A, with a fixed-price option, provided cost certainty while allowing transparency of contract changes owing to the activity schedule, which lists and prices various activities, should more or fewer be required.

The NEC3 Option C target contract with activity schedule was used to execute the majority of construction works contracts, particularly those involving a two-stage tender process with partly defined requirements and where the design was not fully complete (at RIBA stage C or D). One of the main advantages of using an option C target form of contract is that it provides the client and contractor with an actual cost mechanism that can be used to share risk and incentivise performance. Option C can include incentives for the contractor and the supply chain to meet specified targets around time, cost and other predetermined parameters. These incentives may include incentive payments, bonuses or pain/gain sharing of expenditure above or below the original target price. Target prices are based on an agreed activity schedule, which enables the close monitoring of contracted targets against actual costs and programme. NEC3 Option C contracts also allow the contractor the flexibility to reduce costs and other aspects of the programme as work proceeds to achieve incentivised targets. Achieving these targets benefited both the ODA and the supply chain. Option C allows for a design-and-build approach, which transfers the risk of the design activities and the programme risk to the contractor. With appropriate incentives, this can lead to benefits to both parties. Some programmes, such as Crossrail, have used optimised contractor involvement (OCI) or, similarly, the Highways Agency uses early contractor involvement (ECI), which can equally well be used under the Option C approach. OCI and ECI are methods of gaining early inputs or technical advice from contractors.

NEC3 Option E is a cost-reimbursable contract. This form of contract is used only where risk cannot be adequately determined or mitigated by the client. In the event, the NEC3 cost-reimbursable option was never used during the London 2012 programme.

NEC3 Option F is a management contract and was used only in certain instances on London 2012, where packages of works were agreed as the works proceeded. There was also potential for its use where mixed types of structure were to be built in a temporary or permanent environment, such as at the Eton Manor facility, where there was a requirement for the close management of interface with the London Organising Committee for the Olympic Games (LOCOG), which was responsible for the Olympic overlay or temporary facilities.

Finally, the NEC3 Term Service Contract was used where there was a requirement to maintain completed assets through to the completion of the deconstruction period. Certain aspects of this form of contract were included within the main venues, where the contractor retained an obligation to provide a facilities management service. The term 'service contract' allowed the use of key performance indicators to ensure service levels were delivered. At the same time the NEC3 Term Service Contract was flexible enough to allow procurement of additional capital works, if required.

Classification of contracts

The contract classification structure was designed to identify each of the main types of contract to be applied, depending on circumstances. The structure set out a standard solution for each situation, based on the amount of information available to bidders in terms of design and scope, and taking into account the level of risk and the factors of time, quality and cost. Figure 4.7 shows the level of risk on the horizontal axis and the degree of clarity of the expected output, referred to as the definition fixity of the works, on the vertical axis. Where the risk is seen as high and the expected outcome possibly variable, then the appropriate contract category is given as Class A. The main purpose of classifying contracts in this way is to pass on the appropriate level of risk to the most appropriate and best-placed organisation to manage it.

Because of the number of different contracts used in the London 2012 Games, they were classified according to the type of work. Six classifications were used, ranging from Class A to Class F, and each classification was aligned with one or more of the packaging clusters, as follows:

- Class A contracts for permanent structures and associated works;
- Class B contracts for temporary structures and facilities;
- Class C contracts for civil-engineering works;
- Class D contracts for enabling an landscaping works;
- Class E contracts for utilities work, and
- Class F contracts for direct supply of goods and material.

Class A contracts were used for permanent structures and associated works, where the works could be performed within discrete zones – for example, the Aquatics Centre and Velodrome. As the contracts had to deal with complex design and delivery issues, the contractors needed to liaise with the London Organising Committee of the Olympic and Para-lympic Games (LOCOG) and accept high levels of risk in terms of time and cost.

The works were procured using the Public Contracts Regulations (PCR) 2006 restricted procurement route and a two-stage tendering process. The restricted procurement procedure involves limiting the number of contractors submitting tenders to those deemed suitable on the basis of a number of criteria, including financial strength, techno-logical competence and skills. In a two-stage tendering process, the first-stage selection is determined on the basis of the tenderer's proposed construction programme, their method statement and costings. At the end of the first stage a preferred contractor is appointed. The second stage then focuses on the appointment of subcontractors through dis-cussions between the employer and the preferred tier 1 contractor to agree a lump sum for the contract. At the end of the first stage of the tender process the London 2012 programme managers planned

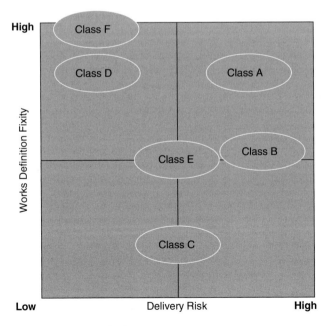

Figure 4.7 Definition versus Risk matrix.

to award the Class A contracts, which were specifically, NEC3 Option C (Target Contract with Activity Schedule) contracts with a gross maximum or target price. This was seen as the most appropriate contract for this particular risk profile. However, where a restricted route was not possible, a competitive dialogue route was used during the stage one tender process. This occurred only on the Aquatics Centre project.

Class B contracts were used for temporary structures and facilities, including basketball, BMX, hockey, Eton Manor and the construction training facility. These were all projects where the scope of the works required relocation or complete removal during the deconstruction phase, post-Olympics, and where the temporary nature of the construction varied from temporary standard components to permanent, but easily demolished, structures. These works also had considerable LOCOG interfaces. LOCOG refers to the London Organising Committee of the Olympic and Paralympic Games, who were responsible for the overall organisation and staging of the event.

The contracting strategy for temporary venues and structures was one of multiple element or trade package project management, and not through a single contract. These assets were delivered through a set of large lump-sum contracts called 'priced works packages'. An example was a contract for the 'temporary pools', which formed one element of the Eton Manor training complex, completed in conjunction with a series of other contract elements including ground works, M&E, temporary structures and building envelope. Some of the contractors bidding for these works were more aligned to the events-management market sector rather than to traditional construction. A mixture of contract types was used according to the risk profile, ranging from NEC3 Option A to bespoke lease-back arrangements.

Class C contracts were used for civil-engineering works, such as structures, bridges and highways, where the scope of the works could be more easily defined. Design and delivery was seen as repetitive and works occurred on multiple occasions and therefore presented only a medium level of risk. A Public Contracts Regulations, PCR 2006 restricted, single-stage, design-and-construct route was used, based on the NEC 3 Option A, Priced Contract with Activity Schedule. However, for a limited number of more complex structures, such as the land bridges, a two-stage route using NEC3 Option C was used. Again, the risk profiles were relatively well defined.

Class D contracts were applied to enabling and landscaping works, including hard and soft landscaping and the North and South Drop-Off points in the Olympic Park. As enabling and landscaping work packages require client-led design, delivery required a high degree of flexibility on the part of the contractors in terms of timing, and because work occurred on multiple occasions. This approach was consistent with the usual strategy for procuring enabling works, which is based on a management contracting tiered framework approach, with the NEC suite of contracts being used throughout the supply chain. The framework applies to a given combination of firms in the supply chain working on tiered projects of varying size and complexity.

Class E contracts for utilities works, such as the combined cooling heat and power Energy Centre and the gas, water and electrical utility networks, required utility sector-specific contracts for design and delivery. The majority of these works came under the Utilities Contracts Regulations 2006 and were contracted through concession agreements, bespoke contracts or uncontested routes. The contested works were procured using combinations of NEC3 Options A and C.

Class F contracts were used for goods and materials that were seen as being of strategic importance for the programme. Class F contracts were supply-only contracts for the direct supply of goods and materials. They were divided into three types of goods, namely:

Class F1 were goods needed as reserve capacity or stock
Class F2 were goods bought and issued free of charge, and
Class F3 for goods purchased and charged.

All other goods were procured with a bespoke ODA purchasing agreement or on sector-recognised terms, depending on the nature of the requirement. Both the clustering structure and the contracting classifications are combined in Figure 4.8, which shows the clusters and their corresponding contract types used to procure packages or projects within each cluster. This combined model of clusters and contracts was used to devise a contract matrix, which was in turn used to determine the number and type of contract required.

Concluding remarks

The packaging and contracting phase of PSE is of primary importance to the success of any capital works programme. It establishes the boundaries and lays the foundations for delivery. PSE views careful planning

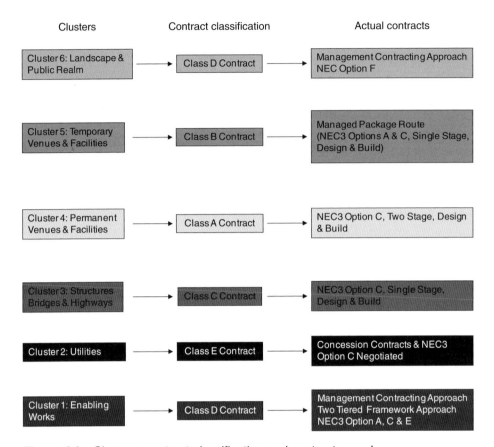

Figure 4.8 Clusters, contract classification and contracts used.

and preparation at the very beginning to be critical for the success in managing a programme. The initial mobilisation phase of a large programme (which was achieved in 100 days by the ODA's delivery partner) should be used to put in place a robust framework, from which a programme approach to procurement can be effectively delivered.

At each stage of planning, the objectives, options, risks and value-for-money requirements need to be balanced to devise an effective procurement strategy. The balanced scorecard can be used to identify and then communicate requirements.

These arise from clear programme objectives in the procurement policy, which are then used to derive the procurement strategy. From the strategy, specific procurement package plans can be developed to select the most appropriate contract options, taking into account both the contracting strategy and associated risk profiles. This chapter has shown how packaging and contracting strategies can be used to procure construction projects within a programme through the use of a systematic approach to the types of contract used. The following chapter examines various approaches to the purchase of common components and materials used in the construction of different projects.

Packaging Strategy

Dos	Don'ts
The most important thing to remember is the strict order of strategy development: packaging FIRST, contracting SECOND	
Identify the full scope of the requirement (project, portfolio, programme, etc.)	Consider the contracting strategy prior to the packaging strategy
Focus upon and understand interfaces and associated risks	Get bogged down in the detail
Understand and consider delivery programme	Over-complicate the strategy
Consider market appetite and capacity through initial desktop study and later via vendor engagement activity	
Determine the organisation structure required to deliver	
Consider the client's risk appetite	
Consider **ALL** services, works and supplies required	
Package to maximise client leverage in the supply chain (e.g. aggregate demand)	
Account for complexity, regularity, value and duration of the scope	
Consider common components and services	
Consider regional/geographic factors	
Identify specialist areas of work	
Ensure that the packaging is reflected in the work breakdown structure (WBS)	

Contracting Strategy	
Dos	**Don'ts**
Consider the macro- and micro-economy	Fail to consider the contractual impact on the lower-tier supply chain
Determine client's appetite for risk	Recommend the contract that you are most familiar with, rather than the most appropriate
Establish the supply market's appetite for risk	Recommend the latest trend or try to second-guess what your client wants to hear
Establish client's requirement for control	
Test if client is an 'expert' or 'lay' client	
Use a complementary suite of contracts, across and down supply chain	
Use legal advice where appropriate	
Provide contract administration training	
Consider the size and availability of the client's support team	
Consider the requirement for incentivisation	
Take account of the Balanced Scorecard	

References

Cornelius, M., Fernau, J., Dickinson, P. and Stuart, M., (2011) Delivering London 2012: Procurement, in Proceedings of the Institution of Civil Engineers, *Civil Engineering*, Vol. 164(5), pp. 34–39.

Ellis, W.D., (1938) *A Source Book of Gestalt Psychology*, New York, Harcourt, Brace and Co.

Palmer, S., (1999) *Vision Science: Photons to Phenomenology*, Cambridge, MA, MIT Press.

Public Contracts Regulations (2006) Statutory Instruments, 2006 No. 5, Public Procurement, England and Wales and Northern Ireland. http://www.legislation.gov.uk/uksi/2006/5/pdfs/uksi_20060005_en.pdf (accessed 1 Sept. 2012).

5

Common component and commodity strategies

Some of the many pathways that now run along the once clogged canals of the Olympic Park (photo courtesy of AECOM).

Programme Procurement in Construction: Learning from London 2012, First Edition.
John M. Mead and Stephen Gruneberg.
© 2013 John Wiley & Sons, Ltd. Published 2013 by John Wiley & Sons, Ltd.

The Velodrome timber cladding during installation (photo courtesy of Mark Lythaby).

Introduction

In the previous chapter the packaging and contracting strategies were discussed in terms of construction clusters. Often overlooked, when considering packaging strategies, is the difficult question of the supply packaging of common components and materials used in different projects and areas of a programme. At the strategic mobilisation stage of procurement for capital works programmes, the common components question is often ignored only to emerge later, by which time an opportunity has been missed. This has been the case on both the London 2012 and Crossrail programmes to a greater or lesser extent.

Two visitors to the Olympic Park record their own images of a memorable and successful Games (photo courtesy of AECOM).

The advantage of the Purchase and Supplier Engineering (PSE) approach is that these challenges are analysed and tested at the earliest stages of a programme, but this does not necessarily mean that missed opportunities will not occur. Missed opportunities can occur for a number of reasons, including the personalities of decision makers, conflicting views, and cultural approaches to innovative procurement that often resist change. While PSE can present the business case, decisions are also influenced, rightly or wrongly, by lobbying within the executive approval process by different stakeholders. However, the use of PSE does make it more difficult for embedded organisational cultural practice to counter sound business case solutions.

The benefits of a common component strategy

Packaging strategies do not apply only to buildings and structures, but can also be used to procure common components and commodities across a large programme or portfolio of projects. The benefits of procuring certain components in a coordinated and uniform way include consistency of design, quality, performance, cost and delivery. Other advantages include improving the security of supply or, conversely, reducing supply risk. This strategy also gives purchasers knowledge of forward demand and enables them to adopt a hedging approach to secure supply well in advance and agree its price. There can also be significant benefits through bulk purchasing and economies of scale, which enable the client to achieve reductions in material prices that, over the extended duration of large programmes, can also help to mitigate inflationary pressures to some extent.

Other advantages of using common components across a programme include the opportunity to solve certain supply problems, such as quality

---

Case study – Crossrail and London Underground escalators

An Industry Standards Group (2012) report recently highlighted the deployment of a Common Components strategy for the Crossrail programme, which could also be of benefit to London Underground (LU) in future. The report stated that:

> The Crossrail project offered LU the opportunity to increase savings compared with the original bespoke proprietary LU escalator design. The Crossrail procurement team included key personnel from the Olympics, where common purchasing arrangements to clear standards had radically reduced costs, increased innovation and delivered successfully. As with the Olympics, getting early and sustained interest from the supply chain was seen as crucial in cutting costs and risks. With 57 escalators needed – approximately the equivalent to seven years of normal LU purchases – there was sufficient size to get a wide range of suppliers interested, particularly as the design was 90% standard . . . with bespoke additions.
> 'The benefits to Crossrail were not just on purchase costs but included reduced civils cost, as the escalators were smaller in cross section so needed smaller-diameter tunnels; reduced risks on handover through early commissioning; 30 years of maintenance from the supplier; and performance level in terms of availability and mean time between failures that exceeded existing LU escalator performance. The target cost for the Crossrail team was 25% less than the lowest equivalent LU escalator and the team actually achieved better than 30% savings.' (Industry Standards Group, pp. 19–20)

control, the logistical flow of those goods to the point of need, a reduction in the number and size of storage requirements, and waste reduction. The use of common components and materials also allows the possibility of achieving consistency of design across the programme for certain elements. Using common components can create an aesthetic unity and a consistent quality across different buildings and structures. This is often required across a portfolio of projects, where a client's corporate identity may need to be reflected in the architecture as a condition of the brief. Common components also simplify and make the facilities management and maintenance of the completed programme much more efficient during its whole life cycle.

Factors influencing the procurement of common components

A number of factors determine whether or not procuring common components at a strategic procurement level is helpful. These factors include market leverage, supply chain security, future maintenance, and operations and design efficiencies.

Market leverage

According to Martin and Grbac (2003), market leverage is one of the benefits of manipulating the supply chain to increase profitability, a method of gearing up of sales and turnover by using the credit of suppliers. Leverage can also give the client one or more of the following advantages. It allows for bulk supply leading to economies of scale and the potential for discounts to be obtained. Because of secured quantities with planned delivery, manufactured products may be improved. Leverage also helps to secure the supply chain for the programme, an especially important consideration when components are critical or their delivery is on the critical path. It also reduces the risk associated with late delivery by securing lead-in times. Finally, by developing relationships with the supply chain, innovation may be encouraged.

An alternative interpretation of market leverage is that it is a measure of a firm's negotiating strength or power, which is a function of the share of a market held by a firm either on the supply side or the demand side. For example, the demand created by one programme may form a significant percentage of the output of the targeted sector or specific supplier firms.

If this demand is fragmented between many separate contractors, prices can rise as shortages appear. However, by consolidating the same amount of demand into one purchase order, suppliers may feel obliged to offer discounts to the purchaser rather than lose the sale to their competitors. Indeed, it may well be possible to play one supplier off against the others to reduce the price charged. This is a typical situation, where the market is monopsonistic and a large single purchaser is in a position to drive down prices. This is the effect of market leverage or negotiating strength. However, if the aggregated demand exceeds the capacity of firms, it can overheat demand and cause prices to rise or delays to the programme. Market soundings and engagement exercises should reveal these issues and make the purchaser aware of the possibility of companies exaggerating their capacity and capability to meet a large increase in demand.

Although the ODA was a client delivering a large infrastructure programme and the leverage may have been considerable, the bargaining strength of the ODA was not necessarily entirely based on the size of its demand. When bidding for construction work, contractors would also have considered a number of other factors in addition to the standard factors of profit and turnover. These other factors include prestige, marketing opportunities, reputation and existing project portfolios.

The client's status in the market is also determined by the programme demand it creates in relation to the overall sector capacity and whether that demand is continuous over a period of years or is confined to a discrete period and represents a one-off demand. Some component quantities may appear large in a programme sense, but taking the construction sector as a whole, or even within a region, these quantities may be relatively small. For example, Network Rail has a high level of demand for steel rail and has continued to do so over many years, while the ODA exists as an entity only until its objectives are achieved, at which point it ceases to trade and no longer creates any demand. The market context determines the leverage of the client as much as do the size and importance of the client.

In addition, where common subcontractors or suppliers are engaged by a number of tier 1 contractors on separate projects, it is important that the client has visibility of the supply chain's capacity across these separate projects within their programme, in order to avoid overextending the demand facing any one tier 2 contractor. Generally, tier 1 contractors will not necessarily have detailed knowledge of supply chains, other than those directly engaged in the delivery of their own projects. Therefore, careful monitoring by the programme client or their delivery partner (DP) is required to ensure that bottlenecks do not develop amongst the lower tiers of contractors, which can have serious knock-on effects on other projects, contractors and the overall programme delivery. The programme client also has an opportunity to monitor many aspects of the delivery of common components, including ensuring that design, manufacture, production, delivery, installation and commissioning are secured across the programme. With their overarching view of the programme, the client or DP may be required to intervene, for example, if tier 1 contractors attempt to deliver their projects without regard to other projects in the programme – sometimes referred to as a 'silo mentality'. Intervention may be needed when there is a risk of excess demand for a common product coming from different projects, or when tier 2 suppliers have insufficient capacity to meet their commitments to different tier 1 contractors. To counter the silo approach to project delivery, there is a need for a strategic view across the whole programme.

Supply chain security

A common component procurement strategy can help to anticipate potential problems in the supply of those components, such as extended supply lead times and competition from other purchasers. Early intervention to buy inputs or reserve supplier capacity can help to avoid unexpected cost increases and reduce the risk of delay. This can also lead to more accurate forecasting and planning.

Future maintenance and operations

Common component strategies can also facilitate reductions in future maintenance costs and efficient operation, for a number of reasons. For example, whole-life costs can be predicted and managed for each type of component. A standard approach to testing and commissioning particular components can be used on multiple projects, taking advantage of the learning curve applied repeatedly to the same product. Familiarity in the use of a component also helps in operations – for example, where building management systems and software can be the same at all locations. Common components also help to standardise maintenance across the programme, thereby reducing operational costs and enabling the use of a single toolbox and one operating manual for each component. Planned maintenance cycles also become easier to manage with a familiar product catalogue across different built assets. The requirement for spares and their management is less onerous than when multiple types of the same functional component are used. With components in common, only a single standard set and reduced number of spares need to be held in reserve to meet the requirements of multiple facilities.

Design efficiencies

Commonality of design helps to ensure requirements are met across the programme by introducing standards that comply with the client's objectives and the practical implications of statutory regulations. This controls the consistency of products through the use of a common specification. By repeating the same approach to design across the programme, there are reductions in the architectural inputs required for detailing. Fewer approvals are required, as the design is for one component rather than multiple components. Similarly, testing requirements (for example, fire testing) are reduced for the same reason.

Developing a common component strategy

There are three basic steps to developing a common component strategy. The first step involves defining the programme scope and listing precisely what needs to be purchased. Consideration needs to be given to select those items in the list of goods and materials that may be appropriate for a particular strategic procurement approach. For example, performance criteria set for a component may determine which components might or might not be appropriate to be bought in bulk or across the programme for different projects.

The second step is to develop a business case to establish whether or not it is worth considering the common elements separately or on a fragmentary, *ad hoc* basis. This is the business case assessment, and justification is needed for each component in order to progress to the next step. The third step is to devise procurement strategies that establish how the purchase of common components can be realised. These three steps for establishing a common component strategy are summarised in Table 5.1.

Table 5.1 Programme scope and procurement

Steps	The tasks
Step 1 – the 'what': performance criteria	Understand the programme scope and specify the products or components to be procured.
Step 2 – the 'why': assessment for common component procurement	Prepare a business case to demonstrate the benefits and test the validity of a fragmented separate procurement approach, taking into account: – the need for consistency, quality and aesthetics – the budget, cost control and cost certainty – risk reduction and risk transfer – relationship to the rest of the programme – liability issues, and – market interest and the capacity of firms to deliver.
Step 3 – the 'how': procurement strategy	Prepare a detailed procurement strategy for each common component to define the delivery and supply model, both before and after purchase. Four options need to be considered. They are: – procure and sell – procure and free issue – reserve capacity and stock holding – joint purchasing for different elements of the programme, where common components can be sourced, as in buying clubs.

Because a common component or commodities strategy involves combining the requirements of many disparate materials and components over a large number of different and diverse projects, it is extremely important to pay attention to detail. Even only slight issues with a poorly executed common components strategy can have catastrophic consequences far greater than the saving in costs of a particular component. For this reason the benefits can be great, but the consequences of error can be serious and therefore the strategy should be developed in three stages, to reduce the risk of making a false economy when purchasing a common component or material.

Stage 1: Performance criteria

The key question is: what are the specifications that the component must meet in order to be acceptable? This critical test establishes whether sufficient and robust component criteria can be generated to communicate the needs of the project teams, taking into account the constraints of time and design aesthetics. This stage therefore engages with stakeholders of the component, (for example, designers and engineers), to establish the criteria to be used for utilising the component. These criteria cover the needs of the various projects, while considering each component's specific market conditions, such as lead times, other competing projects, quality issues and the opportunity to further respond and push the boundaries of design and innovation.

Stage 2: Assessment of benefits of implementing a common component strategy

An assessment of the marketplace is needed to determine the feasibility and establish the benefits to be gained from a common component procurement strategy. This determines the viability of the case for pursuing a common strategy – or not. Those involved at this stage are representatives of the design, procurement, supply chain management and sustainability functions of the programme and any other stakeholders across the programme who would benefit from the implementation of a common component strategy.

This process promotes engagement with the marketplace through channels already established by the programme supply chain management function. This engagement with the supply market assists the team charged with reviewing the business case for this strategy with an opportunity to gain a direct understanding of the engineering, environmental implications, cost, time to supply and visual performance of the components from the supply chain and the market conditions affecting particular components or materials. The business case for implementing the common component strategy must also be assessed for consistency in terms of the delivery of the priority themes established in the programme's balanced scorecard, arising out of the requirements developed at the start of the programme (as described in Chapter 3).

The benefits found and assessed through discussions with firms and market engagement can be set up as a benefits matrix, with the range of common component solutions to be compared and evaluated against the design, engineering, cost and time assessment criteria as set out by the key stakeholders, including the client, end users and investors.

Five tests can be used to determine whether or not to use a common component procurement strategy. The test questions are:

1. Does demand represent a significant proportion of supply and would this gain the client the advantage of leverage?

2. Is there potential to realise cost benefits from implementing a common components procurement strategy?
3. Are there potential schedule benefits from securing early, consistent and guaranteed supply?
4. Are there any potential benefits in the common maintenance of the components during the life cycle of the facility?
5. Are there any overriding design requirements that might prevent common supply?

The answers to these five questions should enable drawing the conclusion as to whether or not a common performance specification alone would be more suitable than a common component procurement strategy. Whatever the result, it would inform the briefing process and, in cases where no action was required, the assessment exercise would nevertheless provide useful information and data. That information could then be shared with design teams and other functions in the delivery of their services.

Stage 3: The benefits of a common component purchasing strategy

A common component purchasing strategy does not increase or decrease the total demand for building components or materials compared with the same materials being purchased separately by different contractors. However, a number of factors determine whether or not procuring common elements on a strategic procurement basis could be helpful. Firstly, owing to the size of very large programmes, if the level of demand created by consolidating expenditure forms a significant percentage of the output of a targeted sector or specific supplier through the purchase by one organisation, leverage can be achieved to the benefit of the purchaser. However, regardless of a common component purchasing strategy, if the aggregate demand from a large programme, together with demand coming from quite separate projects outside the programme, exceeds the capacity of the market, then total demand may cause the market to overheat and prices to rise or production lead times to extend, causing delays in supply.

Under these circumstances the client's programme delivery team may well need to intervene to secure and coordinate the flow of common components and ensure individual projects within a programme do not interfere with each other by seeking to deliver their own projects in isolation, regardless of their impact on the rest of the programme. This is a fundamental problem that needs to be addressed.

Because of the silo mentality often seen in construction and other industries, (Egbu, 2006, and Towill, 2003), programmes may be viewed by managers as a project management solution multiplied by the number of projects in the programme. This approach is fundamentally flawed. The strategic view of multiple projects requires a higher level of vision over the interactions and interdependencies of firms working on the different projects in a programme. This cross-programme horizontal vision or bird's-eye view of the programme is necessary to replace the silo mentality adopted by some managers resulting in projects being built in isolation, regardless of the impact each one may have on the others. A programme is more than the sum of its parts.

Applying gestalt theory loosely to construction programme management, a common component strategy looks for linkages that may not be immediately apparent. The strategy can then be used to provide shared benefits in terms of schedule, cost and logistics. Also, by anticipating potential problems in supply, such as extended supply lead times and competition from other purchasers, early intervention to buy inputs or reserve supplier capacity can help to reduce the risk of delay and avoid unanticipated cost increases.

Using common components strategies to achieve commonality across a programme can also potentially reduce future maintenance costs, for a number of reasons. For example, a store of common parts would help to standardise maintenance procedures and reduce the number and variety of parts held in stock. Common components tend to reduce maintenance cycles through the leveraged purchase of better-quality components, made affordable through the discounts of bulk buying. The common component approach also takes advantage of economies of scale by reducing the management costs of testing and commissioning relatively small batches of components. Finally, a common component procurement strategy can assist with compliance with certain design requirements, including minimum standards and the use of particular materials to meet certain building regulations.

The common component procurement strategy

The common component procurement strategy is the document that sets out the approach to procuring the designated common components, including the different options for delivery. The procurement may be by either the client or the programme manager, directly or indirectly by influencing tier 1 contractors to collaborate, using the market intelligence data gathered during Stage 2 – described above – when assessing the benefits of the common component strategy.

The common component strategy is drafted by the procurement and programme supply chain management team and issued for approval to the programme management executive and client sponsor representatives. The overall contracting strategy ultimately determines the common component procurement strategy, as the latter comes within the former.

A number of basic strategic options were given in Table 5.1 above, including buy and free issue, in which the client purchases the component or commodity and issues it to the tier 1 contractor for use in their delivery. An alternative approach is to use a client supply framework, where the client procures a framework of suppliers, from whom all tier 1 contractors are contracted to purchase specific materials.

A similar arrangement is the tier 1 contractor supply framework, but in this case the tier 1 contractors enter into the contract to purchase the component or commodity they need in common. This creates a contractual arrangement between the tier 1 contractor and the supplier in a separate transaction agreement. This approach was used by Transport for London (TfL) in their escalator framework contract (see case study earlier in this chapter).

Another approach is the benchmark framework, which uses the shared information found in the common component exercise to establish a benchmark of performance. If the benchmark cannot be bettered by a tier 1 contractor's own supply chain, then the contractor must use the suppliers identified during framework procurement. This approach requires additional management input but allows flexibility, especially if market conditions change in the course of the programme, or if a tier 1 contractor's own buying power across their business is greater than that of the programme.

The benchmarks are found by taking various measures from the tier 1 contractors' existing supply base. The tier 1 contractor that has the best value arrangement with their supplier then buys on behalf of the programme. This benefits the tier 1 firms, as they increase their turnover and margins, while gaining from their mark-up to other tier 1 contractors.

Yet another option is the use of a buying club. They can be used where the client is not expected to intervene directly in the supply chain of a common component. This option requires that tier 1 contractors cooperate to aggregate their demand. Their cooperation may be encouraged through the use of incentives or by identifying a direct saving, which can be shared by all participants. However, problems with this approach include possible conflicting objectives, which can lead to unintended consequences. Buying clubs may even require greater coordination and intervention from the programme team than the other approaches. Buying clubs are usually the only option when common buying strategies are being retrofitted into existing contracts, and therefore they require a very strong business case that offers benefits to all.

Concluding remarks

Whatever common component strategy is adopted, it needs to detail the scope, take into consideration the outputs from the market analysis, identify risk factors, assess the various available options for procurement, highlight the value for money considerations and make appropriate contract option recommendations. The form the strategy takes is determined by internal governance processes, external constraints – such as legislation and EU regulations – and attitudes to risk. If an indirect method of delivery is considered appropriate, where delivery is subcontracted to the tier 1 suppliers, then procedures need to be set up that facilitate and foster the necessary cooperation between all supply chain parties, including the client, the tier 1 contractors and the supply market.

The dos and don'ts of common component and commodity strategies are similar to those of the packaging strategy as shown at the end of Chapter 4.

References

Egbu, C., (2006) 'Knowledge production and capabilities – their importance and challenges for construction organisations in China', *Journal of Technology Management in China*, Vol. 1(3), pp. 304–321.

Industry Standards Group, (2012) *Specifying Successful Standards: Infrastructure Cost Review*, London. http://www.ice.org.uk/getattachment/3b96c4c3-9045-4ba0-9549-1586c90019fb/Specifying-Successful-Standards.aspx (accessed 18 Sept. 2012).

Martin, J.H. and Grbac, B., (2003) 'Using supply chain management to leverage a firm's market orientation', *Industrial Marketing Management*, Vol. 32(1), pp. 25–38.

Towill, D.R., (2003) 'Construction and the time compression paradigm', *Construction Management and Economics*, Vol. 21(6), pp. 581–591.

6

Engaging with suppliers: How to attract suppliers and increase interest and awareness

One of the many places to sit, eat and enjoy the Park's surroundings during the Games (photo courtesy of AECOM).

Programme Procurement in Construction: Learning from London 2012, First Edition.
John M. Mead and Stephen Gruneberg.
© 2013 John Wiley & Sons, Ltd. Published 2013 by John Wiley & Sons, Ltd.

With the Velodrome track installed, it is placed under protective wrapping in preparation for the Games (photo courtesy of Mark Lythaby).

The spectacular 160 m long, 3000 tonne steel framework of the Aquatics Centre's single-span wave-like roof takes shape (photo courtesy of Mark Lythaby).

Introduction

The previous chapters have discussed the process of creating an organised system of contracts and clusters in which contractors and stakeholders can have confidence. There is little point in attracting firms to a programme if a framework or the method of working has not been thought through thoroughly. However, much remains to be discussed with potential suppliers, who need to be drawn into the process. This chapter now describes various approaches used to engage with the construction supply chain by raising awareness, building interest and gathering intelligence and feedback to inform programme planning.

PSE does this by targeting appropriate tier 1 contractors and their supply chains. Engaging with the supply chain also involves ensuring the vendors are aware of, and are given the opportunity to input into, the emerging programme. At the same time, the methods described in this chapter are also designed to establish whether or not these opportunities are of sufficient interest to potential suppliers to warrant their investment in submitting a response to invitations to tender. This interest is their appetite, and their appetite is the key to delivering the client's requirements.

This chapter sets out the methods used to engage the target marketplace, which is predominantly those firms experienced in large capital construction works programmes. The chapter describes how procurement strategies are market-tested and how feedback and opinion are sought to ensure that appetite is maximised.

Value for money is a function of the level of competition within a marketplace. Without competition, the power position in a transaction usually rests with the supplier. If only one supplier were to engage in the procurement as a result of poor appetite, a monopoly situation would emerge and the buyer would have to pay a relatively high price to persuade the contractor to undertake the construction work – a price over which the buyer would have little control. Therefore, competitive tension from organisations that wish to win the work on offer is seen as lowering prices and raising value for money. The level of competition is a function of the appetite of firms to enter into the procurement process, based upon their view of the risks associated with the order, their chances of success and also their capacity to meet the level of demand associated with the opportunity. The market's capacity to meet a client's level of demand is a function of the appetite of the market to submit a response to the client. Therefore appetite for any given opportunity is the starting point and key to delivering competition and value.

Is there enough capacity in the construction sector to meet the demand for London 2012?

This might seem a strange question with the benefit of hindsight, looking back on the success of the London 2012 construction programme. The construction industry very obviously had the capacity to deliver. However, at the time that was by no means certain. In 2006, when the question was first posed, there were real concerns over whether or not a very busy and overheated construction sector could respond to a large amount of demand from a newly emerging client – the ODA – to deliver a number of complex construction projects (mainly sports stadia and arenas) on a contaminated site in the heart of east London, to a schedule that had an absolutely immoveable deadline for completion.

The answer given to this question at the time was:

The question should be not whether there is enough capacity to meet the demand, but whether there is any appetite. If there is no appetite, then there is no capacity.

Capacity is a function of appetite and therefore the greater the appetite, the greater will be the capacity. If appetite was high, then competitive tension would be keen and leverage to achieve value would also be enhanced. It was because of this approach that vendor engagement was developed and delivered for the ODA and formed an important part of removing surprises and achieving value.

Engagement with the supply chain or vendors prior to publishing an opportunity is the first point at which a client can hear the reaction of the marketplace and judge its appetite, allowing the response of firms in turn to influence the emerging procurement strategies and package plans. From the contractors' point of view, as the stakeholders who have to deliver value to the client for the capital expenditure programme on offer, they still have to make a profit at the same time. It is therefore important that they are brought into the way projects are to be procured, so they can see the risks and opportunities associated with the award of a contract.

The packaging strategy for a programme depends on a number of factors. Decisions made regarding packaging can impact on the supply market's appetite to respond to procurement opportunities. However some factors, such as national economic conditions and market dynamics, are beyond the control of any programme – but there are a large number of factors that, if tailored to target a supplier's appetite, can attract participation and as a consequence increase competition between suppliers to ensure that tender submissions reflect the capacity of the supply market and offer value for money, while also motivating firms to perform well.

To achieve successful delivery it is vital that the most capable supply chains are attracted to compete for a place on any programme, and that these suppliers understand the requirements for delivery, including all the factors detailed in the client's balanced scorecard. Other critical success factors for delivery also need to be communicated, such as issues relating to sustainability, innovation, regeneration and legacy. Communicating these values during engagement, and testing suppliers' responses prior to the actual procurement, ensure that the supply chain understands the client's priorities and can therefore prepare its response to them.

The engagement of the supply side prior to publishing an opportunity in the marketplace ensures that surprises during procurement are reduced and the number of responses is increased. Engaging the supply chain in advance of finalising the packaging, contracting and overall project or package procurement strategy for a large programme of works increases the probability that appropriate organisations with the required capabilities are informed, and will be ready to respond when the opportunities are advertised. An informed supply chain can effectively plan their resources to provide a response, compete, and deliver the work packages according to the client's requirements.

Gathering market intelligence

Gathering market intelligence focuses on engaging with and identifying potential suppliers. This process involves collecting and analysing data on the supply and demand profiles of different sectors of the market. Figure 6.1 highlights the PSE approach to engagement. The first step is to make contact with appropriate industry bodies and trade associations. This is closely followed by collecting data and information related to the target suppliers' operations and capabilities. Using this information, the next stage is to identify the key factors affecting supply relative to the demand in question and to take these into account in order to plan and make appropriate and timely recommendations. The cycle

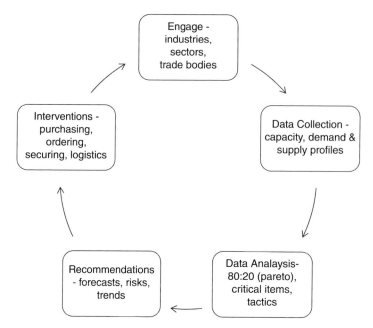

Figure 6.1 The market engagement cycle.

concludes with informed interventions and recommendations being made to ensure that emerging strategies reflect market appetite. The information gathered during this engagement cycle can also be used to identify which firms dominate particular markets and what tactics might be used to motivate these firms to take an active interest in the emerging opportunities. Based on the results of the analysis of data gathered, potential market capacity can be gauged, risks can be highlighted, trends established and high-level forecasts made. The purpose of doing this is to be in a position to make recommendations for any specific actions that might be required during the early engagement phase. These recommendations could include taking decisions to intervene to secure, purchase, order and plan the specific method of delivery.

Throughout the market engagement cycle, those supply chains best placed to deliver the programme are identified and targeted by making them aware of the forthcoming demand of the programme and in particular the client's specific requirements, discussing the priorities in the balanced scorecard. Even at this early stage, prior to publishing any formal tender opportunity, discussions with the supply chain will help to decide whether the proposed programme can realise the client's objectives.

Supplier dialogue

An important feature of PSE is the use of early discussions with suppliers, using a number of different approaches. Each type of approach is designed to gain market intelligence relating to the level of interest and awareness of the programme, while giving an indication of the capacity of the supply market to meet the likely demand of the programme. The dialogue with suppliers takes a number of forms, some of which are one-way supplier dialogues, which promote the programme to the target industry and prepare the marketplace to respond to emerging opportunities to tender. Other dialogues are two-way, which while still promoting the programme are more focused on receiving feedback and gathering information prior to the release of a tender opportunity. These engagement activities can be summarised as:

One-way Supplier Dialogue

1. Supply chain events
2. Use of the Internet – for example, publishing a 'business opportunities' section within the programme's website
3. Publication of a supplier guide.

Two-way Supplier Dialogue

1. Supplier registration and pre-assessment tools
2. Market sounding exercises.

The purpose of entering into a dialogue with suppliers is to ensure that not only are suppliers aware of the forthcoming procurement opportunities, but that they are also party to the development of the emerging procurement, packaging and contracting strategies. One-way dialogue is either the programme presenting information to the supply chain, or the supply chain presenting information to the programme. Two-way supplier dialogues begin with questions from the buyer leading to answers from the supplier, which can be used to draft, refine or finalise policy, strategy and procurement plans.

Figure 6.2   Supplier dialogue mechanisms, their timing and relative cost.

One-way supplier dialogue – Supply chain events

A key element at every stage of the procurement process is the supply chain's need for a human interface at client level for the tier 1 contractors and for the tier 2 and tier 3 levels for the underlying supply chain. Figure 6.2 shows what type of activity can occur, and when, during the procurement process.

One-way supplier dialogue – Industry days

Industry days are usually open events hosted by the programme and are aimed at the supply side of an industry or sector. They are designed to engage suppliers and set out the objectives and procurement requirements for a given project, package or group of procurements. This is a key engagement activity in the very early stages of a programme's delivery, particularly when an early indication of supplier interest is being sought. Such events are especially useful when demand for construction is high, in order to bring the programme to the attention of potential suppliers at an early stage. Industry days also help the wider marketplace to understand the programme, its general scope and context, and both the scale and anticipated routes for procurement. They are also an opportunity for the programme team to set out a high-level timetable. It is usually at this type of event that the balanced scorecard is first presented. This serves to establish the expectations of the programme client by communicating their values and underlying performance criteria early on in the process.

   The use of the balanced scorecard at these early events sets supplier expectations for what exactly is required from the supply chain by the client. Open events like these also facilitate supply chain networking between the various tiers of the supply chain attending the meeting, and provide important initial feedback on the extent to which the procurement strategy has succeeded in creating an appetite for supplier participation.

One way supplier dialogue – One-to-one meetings

One-to-one meetings are used to understand the issues that could prevent contractors from responding positively to specific tender opportunities. The purpose of one-to-one meetings is for clients to listen to the issues raised, to allay fears wherever possible, to encourage appetite and to ensure that a suitable contingency plan is developed, if it becomes clear that there is only a poor appetite in the marketplace for the procurement in its current form. It is common for one-to-one meetings to follow an industry day, as delegates are already to hand. By arranging these in advance of the industry day, key suppliers can be targeted to provide feedback to the programme team and inform the procurement plan.

One-to-one meetings are structured meetings with at least three key members from the targeted organisations invited to be in attendance. These individuals are a combination of commercial, technical and executive level individuals, who can provide feedback on the industry day if such a day had been held prior to the one-to-one meeting. They can also be asked to respond to a standard set of pre-prepared questions. The use of a standard set of pre-prepared questions enables responses to be compared and contrasted with those of other firms. To prepare for the one-to-one meeting, the programme supply chain team can prepare some background intelligence on the participating organisations and this would generally cover the organisation's size, capabilities, current known workload, basic financial information and other relevant areas of interest on the programme.

To remain fair to all eventual respondents to the procurement, the rules of engagement during a one-to-one session must ensure that the programme team ask the questions and the engaged supplier provides answers. Therefore at the start of the session a standard opening statement may be given, along the following lines:

> The purpose of this meeting is for the programme team to obtain direct feedback from the construction industry on their procurement

and contracting strategy and will be based on a set of predetermined questions.

Once these formalities have been covered, the same standard questions can be asked at each meeting. The duration of the one-to-one meetings may vary from 40 minutes to one hour, depending on the nature of the questions and the number of one-to-one meetings to be held. If they are being held after an industry day, the time available is usually restricted and because of the need for a standard protocol the time allocated to each meeting needs to allow for the exiting group to leave and the subsequent group to be brought in. The output from the one-to-one meetings is a summary report to the project team, to be used to inform the emerging procurement plan.

One-way supplier dialogue – Meet the buyer events

'Meet the buyer' events provide a forum where opportunities to improve the value for money obtained from the supply chain are identified through a dialogue between suppliers and buyers. Programmes may arrange their own meet the buyer events for specific common components and commodities, or may attend sector-based events organised by third parties.

At meet the buyer events the programme supply chain team can meet with all interested suppliers and outline the routes to engagement, highlight current opportunities, direct suppliers to any pre-registration mechanisms, give details of the programme's web presence, distribute copies of the supplier guide and generally explain the programme supply chain objectives.

For some large programmes on the scale of London 2012, this type of engagement was particularly effective for resolving logistics issues for timber supply, investigating lighting solutions, sustainable building wrap innovations and supply solutions for programme-wide temporary seating. More recently, on an even larger programme – namely, the Crossrail infrastructure programme – the same approach was used to engage with reinforcement and concrete suppliers to discuss common component supply issues, potential pricing and supply issues and best-value logistics solutions. Feedback from these specific meet the buyer events proved invaluable and consequently shaped the procurement approach for these elements. Meet the buyer events are quite different from meet the contractor events.

One-way supplier dialogue – Meet the contractor events

'Meet the contractor' events serve to pass on the indirect supply chain opportunities that emerge from a programme, from the directly contracted tier 1 contractors to the second-tier businesses. Tier 1 contractors often have large or complex and specialist parts of their supply chain as members of their teams during the development of their own tender. Therefore at these events the opportunities that are on offer for discussion tend to be those of interest to small and medium-sized enterprises willing and able to join existing project teams. The events are the reverse of an exhibition, where sellers hire stands. At a meet the contractor event, buyers are static in stands and suppliers circulate, having prearranged, 10-minute meetings with the tier 1 contractors. This gives sellers an opportunity to have private face-to-face meetings with buyers to explore predetermined specific opportunities that are expected to emerge from the tier 1's own procurement schedule.

To ensure the specific opportunities are valid and that buyers meet sellers who are able to offer appropriate goods or services, the programme supply chain team needs to arrange in advance to have a list of forthcoming opportunities from the tier 1 contractors. Taking this list and working with the relevant trade associations and industry bodies in advance of the event, buyers can be matched to specific suppliers whose services match the emerging requirements. This type of event has been likened to 'speed dating'. Matching supply to demand in this way has proved to be mutually beneficial to buyers, suppliers and the overall programme.

Events like these are an excellent way of giving smaller businesses, with limited marketing resources, access to a number of potential leads, with a specific demand, in a short space of time. They also allow these businesses to gain an understanding of the procurement requirements of the tier 1 contractors awarded contracts across a programme. By working with trade associations, umbrella groups and local business support agencies, the programme team can match tier 1 contractors with tier 2 suppliers, who as members of such organisations are deemed fit to deliver the works, goods and services being identified.

If there is an underlying requirement, set by one of the key stakeholders, to open up opportunities to certain aspects of the wider supply market for political or socio-economic reasons, it may be necessary to write the requirement for tier 1 contractors to participate in this type of event into their contract documentation. For example, it was a requirement of the Crossrail contracts that all tier 1 contractors had to hold a meet the contractor event on an annual basis.

One-way supplier dialogue – Supplier guide

A supplier guide is a short and simple document written and published by the programme to give suppliers an outline of various aspects of the programme and its expected supply chain requirements. It is principally aimed at the lower tiers of the supply chain, but can also be accessed by the larger tier 1 contractors, although they may well already be familiar with the basic content of the supplier guide. The guide is designed to provide useful information to potential suppliers and contractors to ensure that businesses of all sizes understand how emerging programme opportunities would be advertised, competed for, evaluated and won. The document itself may be published both in hard- and soft-copy format and is intended to set out the following general points:

- a high-level general introduction and overview of the programme's scope, scale and associated timeframes;
- the programme's overall vision, objectives and values and specifically how they translate into a balanced scorecard for procurement and delivery;
- the legal framework within which the programme operates and to which it must adhere, including public contracts regulations;
- what and how the programme intends to purchase in terms of goods, works and services;
- how opportunities further down the programme's supply chain are expected to be procured and how those arrangements are to be made accessible to lower-tier organisations;
- an outline of the programme's policies and procedures in general terms, with information about where to obtain full copies of the procedures;
- how interested suppliers can identify appropriate programme-related opportunities, including how local business opportunities may be accessed;
- details of the supplier registration process and how it is to be used to communicate programme engagement and procurement activity; and
- a glossary of key terms that may be specific to the programme, together with links to support organisations and other useful addresses.

Both the Olympic Delivery Authority and Crossrail programmes published supplier guides, which were disseminated to thousands of businesses. During the mobilisation and delivery of the London 2012 programme, supplier guides were distributed at both regional and national business engagement events to encourage interest in the opportunities and encourage businesses to participate in the building programme. Supplier guides are a useful method for any organisation to engage with its marketplace and be transparent about its procurement practices and procedures, ultimately encouraging competition and effective procurement.

The benefit of this passive method of communication in very large programmes is that it reduces the need for direct communication with thousands of individual companies, or at least gives the individual or company making an enquiry a tangible document to take away, digest and, if at all serious about winning work on the programme, consider during the preparation of their response during procurement.

The London 2012 Learning Legacy describes the supplier guide as one of its 'champion products' and describes the benefit to future projects as follows:

[The] guide is presented as a best-practice template for an organisation to engage with its marketplace and be transparent about its procurement practices and procedures, ultimately driving maximum competition and therefore value for money in procurement. (Olympic Delivery Authority, (2011) *Learning Legacy: Lessons Learned from the London 2012 Games construction project*, London, ODA)

One-way supplier dialogue – Business opportunities website

A simple and effective window onto the opportunities of a programme is its programme website. For both London 2012 and Crossrail, business opportunities sections were developed and their contents managed, developed and controlled by their respective programme supply chain teams. This part of the corporate website was dedicated to informing the supply chain of the programme's values, policies and procedures, while linked to the section were the details for emerging procurement opportunities. The business opportunities sections of the websites were used to simplify communication and give access to all levels of the supply chains. The sections also acted as a publicly accessible space, to which all interested suppliers could be directed when they were engaged initially.

The online business opportunities section of a website can be very detailed and the ODA's version evolved over the duration of the programme. In its simplest form, the business opportunities section of a programme website should consist of at least five simple pages, as follows:

- a landing page, which informs the person browsing the internet where they are and outlines what can be found in that specific section of the programme's website;
- a section that covers 'What the programme is buying', briefly setting out in general terms the two categories of procurement related to the programme: namely, direct and indirect supply chain opportunities. Direct procurement opportunities relate to that expenditure spent by the programme client with its contracted tier 1 suppliers. Indirect procurement opportunities are those spent by the tier 1 and lower supply chain tiers in the form of subcontracts;
- a section that sets out 'How the programme will be buying', which contains links to internal or third-party sources for advertising emerging opportunities. For example, on publicly funded programmes it may be appropriate to provide a link to Tenders Electronic Daily. If some form of supplier interest registration is used this can be linked to an e-sourcing tool. This section should also contain a link to the supplier guide to allow users to download an electronic copy;
- a page entitled 'Current Opportunities', which consists of further links to the opportunity slides described below. This allows interested organisations to download copies of these documents and review offline, if necessary. The opportunity slides are owned and managed by the programme supply chain team and should be updated on a regular basis to reflect the most recently approved tender lists and contract awards for tier 1 contractors across the programme;
- a final page on the business opportunities website should contain access to any relevant programme policies and procedures that might assist an interested supplier in understanding the programme's requirements, priorities and overarching policies. This page should equip suppliers with the information they need to provide a comprehensive response to direct contract opportunities when they arise.

One-to-one meetings, supplier guides and the business opportunities section of the website are designed to overcome the possible barriers that inhibit some potentially interested suppliers from becoming involved in programme opportunities. These potential barriers may be caused by a lack of access to current and timely information on direct opportunities, or by misunderstanding the timing of procurements. By publicising a programme's procurements and giving details of their progress as they develop, suppliers can effectively mobilise and respond with quality submissions. The use of IT and the internet provides firms with up-to-date access to enough basic information to enable them to plan and allocate resources ahead of procurement commencement dates.

One-way supplier dialogue – Opportunity slides

London 2012 and Crossrail were keen to ensure the high visibility of information relating to their respective pipelines for procurement across their respective programmes. Opportunity slides were devised as simple documents listing all direct procurements with basic information, which included forecast contract notice publication dates, procurement status, and the progress of procurements during the tender stage. The slides also named short-listed tier 1 contractors, in order that tier 2 organisations could pursue potential supply chain partnerships and market their goods and services directly to the tenderers. Once awarded, the successful tier 1 organisations were listed with their agreed contact details. This enabled the supply chain to make contact to understand the awarded tier 1 contractors' emerging requirements and to explore any opportunities for working with them on the programme. By updating the opportunity slides regularly, the programme supply chain team were able to keep all prospective suppliers up to date on the current status of their procurements.

Opportunity slides are essentially a list of all the key programme procurements, including the status of the procurement, a one-line description of the package, a target date for awarding the contract and an indication of a range of anticipated costs. Suppliers could use the opportunity slides to track opportunities and respond either directly by monitoring the appropriate channels, or indirectly by waiting for tender lists to appear on the slides and then approaching tenderers directly.

Two-way supplier dialogue – Supplier registration and
pre-assessment questionnaires

To ascertain the appetite, capacity and hence capability of the market to supply goods and services, a supplier registration process can be used.

There are various existing third-party solutions that can assist in this process. Some are straightforward electronic brokerage systems, such as CompeteFor, www.competefor.com, a free-to-use (for both buyer and vendor) website that was developed for the purpose of opening up supply chain opportunities arising out of the London 2012 programme. The tool was designed to allow buyers to identify a long list of potential suppliers that matched certain supply requirements. Just as the Official Journal of the European Union (OJEU) advertises contracts being procured by public-sector organisations, CompeteFor does the same but for all manner of opportunities, private or public, large and small alike.

Many of the other third-party systems go beyond the advertisement of opportunities and assist with the procurement process itself. That is to say, they are the starting point for the full procurement process. These systems vary in their complexity and requirements but, unlike the electronic brokerage system developed and used by the ODA, they usually involve the payment of a fee, either from the buyer or the vendor – or in some cases both.

If a programme is large enough, it may justify investment in the development of its very own simple registration system to capture basic information on interested suppliers. This would give suppliers an opportunity to offer their products and services and would provide the programme with valuable data.

In the earliest days of the programme to deliver the infrastructure for London 2012 and before the implementation of CompeteFor, firms were invited to express interest via the London 2012 website, registering their most basic contact details. This data formed a simple email list for communicating to all interested organisations.

As part of a programme's own registration process, a pre-assessment questionnaire (PAQ) can be used to gauge whether a supplier is fit to supply and therefore to compete for a place in the programme supply chain. A PAQ should be very simple, but aim to capture useful high-level data that can be used later. The design should utilise the internet and enable online registration and allow some automatic assessment using closed questions, inviting yes or no answers. A negatively scored answer would therefore highlight a gap in the ability of a supplier to win work. The firm could then automatically be directed or passed to a third-party organisation, such as a Chamber of Commerce or public-sector-funded advisor – for example, BusinessLink in the UK – who could then help the business to acquire the knowledge or skills needed

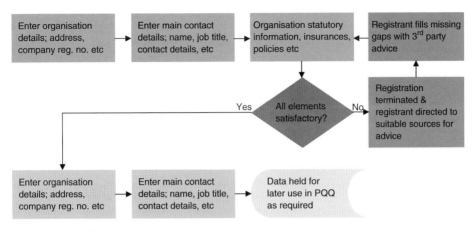

Figure 6.3 Simple pre-assessment process.

to become a 'fit to supply' vendor. Figure 6.3 shows a process map for a basic pre-assessment questionnaire procedure.

Apart from the PAQ, suppliers are asked to supply information relating to their appetite, capacity and capability to supply. This information takes the form of financial data (including turnover) and a breakdown of the business owners and their employees. Particular objectives of the client can be audited at this stage in the selection process. For example, to meet one of the priorities of London 2012 to engage with small and medium-sized enterprises the turnover data could be used to gauge the size of the organisations registering.

Other priority areas might include geopolitical and demographics data relating to diversity of business ownership, their specialisms and what type of industry they operate within. For national programmes of work, interested organisations might also be asked to identify in which geographical areas they would be interested in working. This data collection is the beginning of the compilation of a body of evidence to engage with all levels of the supply chain.

This full registration process has an important and worthwhile side effect. Businesses learn what they need to do in order to fulfil the basic qualification criteria not only for the programme in question, but also for other business opportunities. This not only facilitates the dissemination of opportunities widely to all interested parties, but also leaves a legacy of more developed suppliers fit to supply better goods and services to more wide-ranging clients and future opportunities.

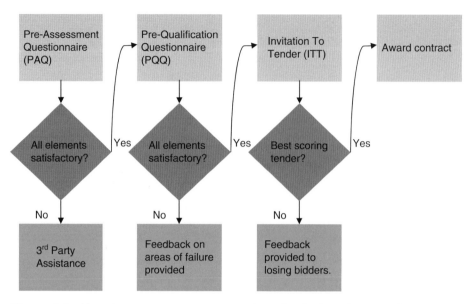

Figure 6.4 How the pre-assessment questionnaire fits into the overall procurement process.

Once registered, when a suitable opportunity with the programme arises the data gathered from the PAQ can be used to complete the most basic elements of the formal pre-qualification questionnaire (PQQ). If the two systems, PAQ and PQQ, are integrated, it demonstrates a seamless and well-delivered, efficient procurement process. Figure 6.4 shows how the PAQ process fits within the standard e-sourcing procurement process.

Two-way supplier dialogue – Market soundings

At any one time, there are always a number of different construction and infrastructure projects and programmes competing for similar resources within a similar timescale, although the amount of total construction work does vary from year to year. Coming to market with good intelligence increases the chances of achieving a programme's objectives and avoiding surprises and disappointment when the tender return day arrives. It is therefore necessary to make contact with potential suppliers through trade associations and umbrella bodies to identify which firms may have sufficient capability and capacity to respond to a particular procurement opportunity and then deliver the works to a satisfactory conclusion.

Market soundings are a method of entering into an early dialogue with supplying firms, and they help to indicate which firms are likely to respond to the procurement opportunities on offer – a measure of the appetite of firms to participate in a programme. The information is useful in case there is a need to adjust or take into consideration market feedback. The mechanism used to gain the feedback is the market sounding exercise. This takes the form of a targeted engagement of vendors with a standard questionnaire and is designed in conjunction with the procurement team responsible for writing the procurement plan and the technical team, who define the scope of the work.

The questionnaire needs to be designed to give the targeted suppliers a high-level view of the project or package of work being procured. It should outline the scope of works, location, interfaces with other aspects of the programme, the proposed form of contract and procurement route, and should also give an indication of budget values, although it is sometimes difficult to get clearance from the client to publish these. However, if they are not included it is often the first question asked by potential respondents to the market sounding exercise. Even giving a wide range of figures is therefore preferable to avoiding the issue altogether.

From the point of view of the client, market intelligence is needed to identify which organisations are available with sufficient capability, capacity and appetite to respond to their opportunity. Market intelligence also helps to assess the probability that a proposed programme can realise its objectives, given the available capacity of firms expressing interest.

The programme supply chain team is charged with identifying a target list of suppliers for participation in the market sounding exercise. They initially engage with relevant umbrella bodies, industry and trade associations. These trade bodies exist to promote the interests of their members and their trade sector and are therefore ideally placed to provide lists of firms with sufficient capacity and capability to undertake the projects or packages being procured. By using these industry organisations it is possible to assemble a list of vendors, who at least profess by their very membership to be suitable players in their sector. Targeting firms using these member organisations also prevents challenge from outside suppliers, who cannot argue that there has been any form of favouritism, as trade associations and similar bodies are most often not-for-profit organisations, existing for the benefit of their members and their sector. This is precisely the reason to engage with them. In every field of commerce there are many business associations and these are usually extremely helpful when seeking to engage with an industry, sector or specialist trade.

The programme supply chain team uses the body of knowledge and intelligence gained from the market sounding exercise to compile a report that informs the procurement plan. A wide-ranging variety of feedback may be forthcoming and it is this intelligence directly from the supply chain that allows the procurement plan to take into consideration all manner of suggestions to avoid risk and increase appetite. It is particularly at these early procurement stages in the delivery of a programme that risks can be avoided. Almost all other supply chain risks materialise once a contract is in place, by which time they are difficult to avoid and can therefore only have their impact mitigated.

Although the feedback and suggestions provided from a market soundings exercise are almost always driven from a partisan perspective, that bias should be ignored, because the suggestions invariably highlight the route to best value. For example, the risks associated with interfaces with other packages can be highlighted from the delivery perspective. It is possible to identify potential risks and opportunities and in doing so take appropriate action in a timely manner by adjusting the procurement plan to reflect these considerations. For example, adjustments to the emerging procurement plan may include one or more of the following areas:

- Packaging – the feedback on packaging is key for consideration. The question is whether or not the scope and scale of the works is too large or too small to generate a strong appetite and attract the 'A' teams of suppliers when it comes to delivery.
- Contracting strategy – do the procurement route and the contract chosen transfer risk in a way such that the risk is placed with those best placed to manage it?
- Bulk purchasing – feedback information can help to decide that a programme has greater leverage of supply than a supplier, or that quality and consistency may be better controlled through client intervention rather than a supplier's own sourcing.
- Specification alignment – feedback can identify areas where specific items might be directly aligned to enable strategic procurement of some common components or commodities.
- Securing manufacturing capacity or output – if demand is high for the goods and services in question, feedback at this stage may indicate specific areas within the procurement where some form of forward buying or hedging may be required beyond the current package.

- Logistics – market soundings may provide evidence that on a particularly large or complex programme the logistics of delivery and distribution across multiple contracts may be an area of concern or may require greater clarity during procurement to avoid bidders including elements of risk pricing.

Market soundings create an early opportunity for firms to be given an understanding of the client's approach to specific projects or package procurements. They also allow suppliers to provide their views on these approaches by giving their feedback. Market soundings can assist firms to consider their approach when compiling their submissions. Upon completion of the engagement exercise with suppliers, key themes and recommendations received in the responses provided can then be fed into the final wording of specific packaging, contracting and procurement. The feedback from the market soundings can also be used to model the capacity of the respondents to deliver future demand and further refine the target list for engagement, once the actual procurement begins.

In order to engage with the marketplace and glean feedback and views on emerging procurement strategies, the first step is to identify the target supply market. This market might be global or local, depending on requirements. It could embrace multiple disciplines and long and complex supply chains, or it might require very specific and specialised skills. It may also look to large multinational suppliers with a wealth of resources to draw upon, or it may require a boutique or bespoke response tailored to a specific need.

As stated earlier, if appetite leads to competition, then the main aim of competition is to leverage the best-value response from within that competition. Knowledge of a target supply market needs to be narrowed down to specific market sectors, trade bodies or even specific suppliers. The aim of engaging a target supply market is to gain specific intelligence from the suppliers operating within it in relation to a number of key questions regarding:

- Appetite – is there interest in responding to the opportunity?
- Capacity – is interest coming from suppliers who have the capacity to deliver?
- Capability – do the interested suppliers have the right capabilities to deliver the specific requirements of the opportunity?
- Demand – what is the extent of current and future demand from other construction projects for similar kinds of work and are we, as clients,

competing with other clients whose model presents a better, or different, set of propositions?

- Potential risks – what do the target suppliers see as the main risks to delivery, and are they comfortable with the way in which these risks are allocated within the procurement strategy?
- Current issues – are there any specific issues within this market that may restrict or delay responses to the procurement opportunity or delivery of the contract?
- Trends – is there a trend in the supply market for a certain form of contracting or delivery model that has not been considered, or that is preferable in order to strengthen the appetite of potential contractors and suppliers?

London 2012 Aquatics Centre case study

When London won the contest in July 2005 to host the Olympics, the initial appetite to bid for sports stadia construction projects was relatively low. This was partly due to the buoyant private-sector market for construction in general, especially speculative office development in London, and partly because the construction sector's track record for stadium delivery was strewn with spectacular delays, cost overruns and losses at that time.

The packaging strategy initially proposed for the Aquatics Centre procurement separated the Aquatics Centre from the main park's entrance bridge, known as F10. The F10 bridge had a direct construction interface with the complicated and challenging structure of the Aquatics Centre. Initial market intelligence had indicated to the ODA and its Delivery Partner that the risk of this complex interface might not be conducive to attracting any contractors, as it had been intended that the proposed packaging was to be managed across two contracts by two different procurements and therefore potentially two different contractors. As a result of feedback on the proposal, the packaging strategy was adjusted to bring the construction of the bridge and the Aquatics Centre into a single and much larger package.

This route had not originally been chosen, as it was felt the package size was too big and the construction disciplines involved in the building of a bridge and an Aquatics Centre required different market sectors to respond. The change of approach was not unique. The type of market engagement utilised by the PSE model and described here has often influenced procurement strategies to the benefit of both sides of a transaction.

Concluding remarks

Supplier engagement is central to the success of a programme. The same can be said to apply to individual large projects that generate large-scale demand for elements of the supply chain. Early engagement can take many forms prior to finalising the supply chain strategy. For one major transport infrastructure scheme in the UK, Crossrail, a number of umbrella bodies and trade and industry organisations were consulted, including the Civil Engineering Contractors' Association (CECA), the Rail Industry Association (RIA), the Construction Products Association (CPA), the Specialist Engineering Contractors' Group (SEC Group), the National Specialist Contractors' Council (NSCC) and the Lifts and Elevators Industry Association (LEIA), to name just a few! This demonstrates the diversity and relevance of the numerous organisations that exist in industry. These organisations are in an excellent position to draw on their members' expertise to provide valuable market feedback and produce data to inform a programme's procurement policy and its supply chain procurement strategies, and for that reason should be engaged.

Close working relationships with those and similar groups demonstrate the commitment of the programme team and their intention to listen to the supply market and engage wherever possible. Communication lines with all these organisations tend to smooth the management of queries and facilitate the dissemination of programme information. Being seen to be fair to all parties and working closely with industry-wide bodies and trade associations helps to gain industry support for the programme in question.

Early engagement of the supply chain also helps to anticipate the responses to emerging procurement opportunities. However, the proof is in the actual responses received. The next chapter describes the PSE approach to standardising programme procurements, or as it has been described 'building a procurement machine', which is efficient and robust but also offers completely transparent and auditable procurement processes.

Vendor Engagement	
Dos	**Don'ts**
Identify and know your target market	Limit communication with the supply chain at any level within their organisation, e.g. use all of your personal contacts to identify/get to the one you want/need
Understand market capacity	Leave yourself open to procurement challenge (public)/criticism (private) through biased communication with suppliers
Understand market capability	Be inconsistent with your message to the marketplace
Develop and implement strategy for engagement	Be afraid to challenge the procurement packaging strategy to ensure maximum market appetite
Utilise trade associations and professional bodies to identify/target suitable suppliers	
Consider the most effective/efficient communication routes, e.g. one-to-one meetings, industry days, road shows, etc.	
Be prepared to meet all capable organisations on an equal basis, e.g. do not show favouritism	
Maintain a single point of contact with each supplier	
Treat all prospective suppliers equitably	
Proactively ensure that the output from vendor engagement is fed back and considered in the evolution of the packaging, contracting and procurement strategies	

Reference

Olympic Delivery Authority, (2011) *The London 2012 Learning Legacy: Lessons learned from the London 2012 Games construction project*, London, ODA. http://learninglegacy.london2012.com/documents/pdfs/procurement-and-supply-chain-management/40-supplier-guide-pscm.pdf (accessed 18 September 2012).

7

eSourcing and process codification: Standardising programme procurements

The Olympic Park gets the first of its 4000 newly planted trees (photo courtesy of Mark Lythaby).

<hr/>

Programme Procurement in Construction: Learning from London 2012, First Edition.
John M. Mead and Stephen Gruneberg.
© 2013 John Wiley & Sons, Ltd. Published 2013 by John Wiley & Sons, Ltd.

The completed Arcellor Mittal Orbit takes its place in the Olympic Park (photo courtesy of AECOM).

Introduction

In the previous chapter various techniques were employed to attract and motivate firms to participate in a construction programme. This chapter describes the actual processes and procedures of procurement and sets out how these processes are set up to ensure consistency and compliance. The Purchase and Supplier Engineering (PSE) approach to the procurement process is to make it systematic and standardised for

The Main Entrance Plaza to the Olympic Park during the Games (photo courtesy of AECOM).

efficiency and transparency. Designing these procedures to meet the requirements of good governance is analogous to building an efficient and robust procurement machine.

With a balanced scorecard, which translates the client's values into meaningful measures and, once clear packaging and contracting strategies are in place with appropriate supplier engagement, the PSE process can embark on the visible procurement activities. Visibility implies formal market engagement with suppliers and the start of the prescribed selection processes. This formal part of the procurement process is subject to legal and third-party scrutiny. It is therefore important that each step of the process is correct in its compliance with all the necessary and associated legislative and governance requirements.

The guiding principles of a robust procurement process

The guiding principles of a compliant and robust formal procurement process include clarity, efficiency, transparency, auditability and accessibility. For example, for clarity of purpose all parties to the procurement need to be clear on (a) what is being procured, (b) how it is being procured, and (c) what the client is seeking to test and evaluate during the procurement process. This involves defining the entire scope for any given procurement and ensuring that the scope encompasses and tests the responding suppliers' ability to deliver what is being asked of them. The aim is to make sure that the time and cost spent in the formal procurement process is kept to an acceptable level for all those involved.

Transparency is another key requirement, especially in public-sector procurement. Transparency demonstrates to all parties involved the openness, fairness and probity of the competitive selection process. Formal procurement processes not only have to be fair, but they also have to be seen as fair. Transparency encourages participation and, as discussed in earlier chapters, raises the degree of competition, which is important for delivering value for money.

Auditability is concerned with the scrutiny of the selection process by third parties. Even when procurement is not being undertaken under formal legislative regulation, it is nevertheless essential that a formal procurement process clearly identifies decision points and provides evidence for the appropriateness of the decisions taken. Formality allows for decisions to be fair and for them to be shown to be so. This requires a clear audit trail for all actions involved in the procurement process, which also gives all parties confidence in the decisions reached and allows buyers to provide comprehensive feedback to those suppliers requesting it at the end of the formal procurement process.

Accessibility is concerned with attracting the most capable supply chains to compete. To complement the vendor engagement work that may have preceded the formal procurement process, it is important to facilitate supply chains to then engage with the tendering process, while retaining the probity of the process itself. This requires removing obstacles that might impede or deter firms from submitting tenders. For example, it is important that all suppliers have equal access to the information and documentation, whether through internet technology or traditional hard copy, and that the documentation should be appropriate for the procurement in question and for the type of firm being encouraged to bid.

Standardising procurement documentation

Standardising the procurement documentation enables the guiding principles to be enshrined in the procurement process and provides standard template documents for the common elements of the procurement process. Standard template documents generate a return through higher productivity, increased efficiency, greater consistency and a reduction in the risk of a procurement failing due to inappropriate activities resulting in a challenge.

The development of standard documentation creates an opportunity to consult those with commercial, technical and legal responsibilities within the client's organisation about their specific requirements from the procurement process. Advice may also be needed on the actual content of the documentation from technical experts – for example, on health and safety issues. The process of developing standard documents may in itself take many weeks, require multiple meetings and a great deal of effort by the parties involved. However, the importance of drafting standard procurement documents should not be underestimated.

Security of the procurement system

On any procurement, but more importantly on any programme of procurement, it is important that the correct information is supplied to the intended recipients and only to them. Procurements require bidders to share confidential information with the buying organisation as part of their tender proposals, which among other things may include intellectual property and other commercially sensitive information. The security of the information being passed via electronic or other means must therefore be guaranteed.

Linked to the need for security is the requirement that individuals be accountable for the information passed between the parties, which in itself acts as a defence against fraud and corruption. It is also imperative for the client to be able to demonstrate that the procurement documentation has been delivered to the responsible people in their respective organisations and that their tender returns have been received and opened at the correct time. On the London 2012 Olympic programme this principle applied not only to the tier 1 contractors, but also to the numerous small and medium-sized enterprises (SMEs) that had expressed an interest in contributing to the programme, either directly or indirectly, as part of the supply chain.

This was especially important for London 2012, as one of the stated aims of government had been to make opportunities arising out of the games infrastructure works available to the entire business population, from large multinationals down to both local and regional SMEs. The procurement process therefore needed to be accessible to all, regardless of size or location, not only in terms of knowing whom to contact but also in terms of the tendering processes and procedures.

Evaluation of tenders

The evaluation of tenders depends on the responses of bidders to questions set in the tender documentation. A significant proportion of these questions are developed as part of the standardisation of the procurement process described earlier. Others are developed to test the specific technical requirements of the contract being procured. To ensure the responses to these questions are evaluated consistently, accurately and fairly, subject-matter experts evaluate the answers and allocate a score and accompanying rationale to each tenderer's response. As with the Balanced Scorecard, questions vary in importance as far as the client and stakeholders are concerned. While some questions may be absolutely critical to the success of a programme, others are optional or only desirable. The evaluation model for each package or project procurement therefore weights each question to reflect its relative importance.

In public-sector procurements, in particular, there is a risk of challenge to the procurement process, which can stop or invalidate the entire process; these challenges can emerge either during the procurement process or, more likely, once an award has been made. Challenges from disgruntled suppliers may be made on the basis of perceived non-compliance with legislation or breaches in the overarching obligation of fair treatment. It is therefore necessary to use subject-matter experts to evaluate the answers provided by tenderers in response to questions posed around their area of expertise. Their evaluations build up the tenderers' scores and the rationales for those scores, before awarding a contract to the offer that represents the most economically advantageous tender (MEAT).

On major programmes it is advisable to provide a separate evaluation management team. The evaluation management team manages the process by compiling and testing evaluations in order to understand and highlight divergences between the subject-matter expert evaluators.

The complexity of the evaluation process on major programmes is caused by the large number of deliverables expected by different stakeholder bodies. The requirement to satisfy all of these bodies at the highest level influences the number of criteria to be tested and evaluated. Having so many criteria tends to dilute the impact of any one of them. The skills required to undertake the evaluations include the ability to assess the appropriate weightings for each area of testing; some of the detailed evaluations may require specialist external evaluators, who are then moderated by the evaluation management team.

The application of electronic tools in the procurement process

The use of information and communication technology (ICT) to transport and communicate documents electronically is far more efficient and cost-effective than producing hard copies, organising their delivery and administering the paper trail they leave. This means that running a formal procurement process electronically is in practice the preferred method of delivering what is needed, as well as offering a far more sustainable outcome in both economic and environmental terms.

> In one year, it was calculated that the Crossrail programme's total use of paper, if stacked in a single pile, would have reached above that of one of the tallest man-made structures, the Burj Khalifa in Dubai (almost 830 m high). And Crossrail was not even using paper-based procurement!

There are, however, a number of constraints when using electronic tools to manage the overall procurement process. The systems have to be secure, yet accessible. Information sent on either a CD-ROM or DVD in essence provides 'electronic paper' to suppliers and therefore retains all the issues associated with communicating physical media between parties. For example, they rely on physical delivery means, which are prone to delay, loss or damage.

The formal procurement process employed in a PSE approach makes use of two very different electronic tools, shown in Figure 7.1. These are electronic sourcing (eSourcing) and electronic evaluation (eEvaluation). The role of the eSourcing tool is to act as the interface between the buying and selling parties, while eEvaluation makes use of the facilities offered by ICT to compare the offers of different vendors. These two electronic systems can be delivered by a single tool, but for ease of explanation they are described separately here.

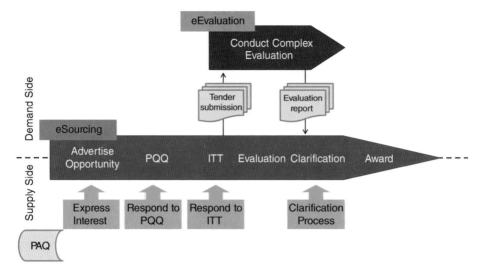

Figure 7.1   The formal procurement process: eEvaluation and eSourcing.

eSourcing

An eSourcing system needs to be accessible to all potential vendors; therefore all commercially available eSourcing tools utilise the internet, and almost all are web-based. Typically, eSourcing systems need to cover the interactions between the buyer and the suppliers at different stages of the process, beginning with the initial publication or advertisement of the opportunity through to the bidding process. Generally, eSourcing works by allocating responding firms a log-in to an advertised opportunity. This grants them access to the online pre-qualification questionnaires (PQQ). If a respondent's PQQ is accepted, or forms one of the highest-scoring returns, they are then sent an invitation to tender (ITT). In the public sector the advertisement of a contract tender opportunity is called a contract notice, and that is published via the internet in the Official Journal of the European Union (OJEU).

The eSourcing tool is used by responding organisations to enter data about their firm and their offer and to submit the PQQ and ITT information. A fully functioning eSourcing system also needs to be able to deal with two-way communication and allow for exchanges of information and a wide variety of issues that flow as part of the buyer–supplier interaction. These interactions range from queries related to the issued documents and issues requiring clarification to the final agreement of the contractual documents.

Reading Figure 7.1 from left to right, the lower section shows the eSourcing interactions between the buyer and suppliers. The process begins with the publication of the opportunity, to which suppliers are invited to provide an expression of interest (EOI). From EOI the process moves on to supplier qualification by way of the Pre-Qualification Questionnaire (PQQ). The PQQ is designed to establish potential suppliers' capabilities, their financial strength and their internal policies relating to subjects such as health and safety, employment and the environment. If acceptable to the client, potential suppliers are then invited to submit a tender for the specific opportunity on offer for detailed evaluation. During evaluation, the system may be used to clarify any outstanding issues before finalising scores and making a recommendation to award the contract.

The eSourcing tool holds all the commercially sensitive information relating to each supplier's response to each procurement opportunity. This means that the system must have sufficient security accreditation to be acceptable to all the parties involved. Access to the eSourcing system is password-controlled on both the client's and the suppliers' sides. To enter the portal, suppliers are directed to a publicly accessible page on the internet. The screen shown in Figure 7.2 is an example of a 'shop window' of an eSourcing system that is provided by the Royal Institution of Chartered Surveyors (RICS).

Only those on the client's and suppliers' sides can therefore see the procurement events, including the PQQs and ITTs, to which they have been given access. The eSourcing tool also needs to be accredited, so that the contractual obligations that are being entered into through the system are legally binding and conform to all applicable legislation and regulatory frameworks.

It is therefore of paramount importance that the security of the system not only gives suppliers confidence that the commercially sensitive information they load onto the system is safe from their competitors, but also that the client cannot view their offer until the appropriate time, as designated by the timetable for procurement. For both the client and responding supplier it is vital that, once the tender is opened, only the client can view all the suppliers' offers, and that suppliers cannot subsequently amend their offers or submissions. This functionality ensures the probity of the process and delivers the auditability needed to run a fair competition, which all parties have a right to expect of a robustly run and properly administered procurement process.

Figure 7.2 The home page of the RICS eTendering site (www.ricsetendering. com). (Reproduced by kind permission of the Royal Institution of Chartered Surveyors.)

Full eSourcing systems need to be able to provide transparency wherever the interactions between the engaging parties are obliged to be audited. This is achieved by ensuring all activities in the system are tracked and logged. For example, if an issued document is updated, the system logs the date and time the new version was uploaded. It also logs whenever each of the suppliers views or downloads a new version of a document. This is important to track changes and place them within a timeframe for compliance requirements.

A fully functioning eSourcing system must include web-form functionality, which enables all forms to be completed online. This allows the PQQ and ITT to provide for specific and precise technical questions relating to particular procurements that suppliers are required to answer.

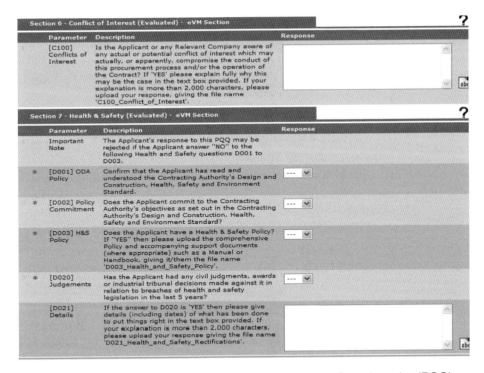

Figure 7.3   Page taken from an online Pre-Qualification Questionnaire (PQQ).

Technical questions can also be designated as mandatory. This level of control over the questions being asked drives the standardisation of submissions and ensures all respondents are fully aware of what is required of them. An example of an online page of a PQQ is given in Figure 7.3.

Certain eSourcing systems allow commercial pricing to be included in web forms. With consistent and comparable pricing structures in standardised tender documents, the online objective assessment of individual items is possible, if structured appropriately. The use of this type of tool requires a full understanding of procurement and the systems involved. An example of a page from an online pricing document is shown in Figure 7.4.

Apart from the online forms described here, an eSourcing system also allows for additional communication between vendors and buyers. For example, vendors are able to post messages on the system, which the buyer's procurement leader receives in the form of an email in their usual inbox. This direct but 'on the record' form of communication means that the system is consistent with the guiding principles given at the start of this chapter.

| Subheading | Earthworks | | | | | | | |
Reference	Description	Unit of Measurement	Quantity	Unit Price	Price	Programme Operation		Comments
* [B411]	Remove protection layer/surcharge (arisings to be removed to on site spoil heap)	£	1					
* [B412]	Excavation for foundations (arisings to be removed to on site spoil heap)	£	1					
* [B413]	Filling	£	1					
* [B414]	Works to formation level	£	1					
* [B415]	Sub base	£	1					
* [B416]	Form landscape areas	£	1					
* [B417]	Verges	£	1					
* [B418]	Reinforced concrete edge upstand / retaining wall	£	1					

Reference	Description						
Subheading	Road Construction						
[B421]	Main Loop road and bellmouths						
Reference	Description	Unit of Measurement	Quantity	Unit Price	Price	Programme Operation	Comments
* [B421a]	- Flexible road base	£	1				
* [B421b]	- Flexible road binder	£	1				
* [B421c]	- Flexible road surfacing	£	1				
* [B421d]	- Kerbs	£	1				

Figure 7.4   Page taken from an online pricing document.

Case study

'Typically the productivity potential of eTendering technology is in the range of 25–50 per cent, depending on the depth of automation functionality utilised by the user. However, the ODA was particularly successful at embedding the use of templates and automation and achieved towards the top end of the range. This significant productivity improvement in effect represented an additional resource pool available to focus on key strategic and tactical objectives.

Between February 2007 and April 2011, the ODA had undertaken more than 580 tender projects involving more than 21,000 suppliers, including about 150 PQQs and about 700 ITTs. This is estimated to have led to:

- productivity savings of more than £5m;
- postage and print savings of more than £350k;
- significant environmental benefits;
- a massive reduction in project cycle time, leading to objectives being delivered faster.'

Quinn (2012)

eEvaluation

A robust and efficient way to deliver multiple evaluations to multiple evaluators is to use an eEvaluation tool. With simple one-off evaluation exercises it is possible to capture subject-matter expert evaluators' rationales and scores on spreadsheets and for the process still to be manageable and efficient. Simple procurements could equally be undertaken using an eEvaluation tool, but this type of tool is particularly useful where procurements are complex, with significant numbers of evaluators, and also where there is an ongoing programme of procurement that requires consistency of evaluation.

The upper part of Figure 7.1 illustrates the role of the eEvaluation function. The eEvaluation tools are distinct from eSourcing tools in that they are used only by the buyer. This means that the eEvaluation tool could, in principle, be hosted on the client's own internal information and communications technology (ICT) network or intranet and within its own firewall. However, in practice remote access and multi-organisation working mean that an internet-hosted, browser-based tool is invariably used. This enables specialist evaluators, who may be geographically dispersed, to access, assess and score the suppliers' submissions by completing their sensitive work in the privacy of their own homes or off-site, where they will not be disturbed. More importantly on large programmes, where co-location is used (with several organisations under one roof), evaluators can be removed from the view of potential suppliers who may be present in the client offices. Alternatively, where offices are shared and teams are integrated, a separate and secure suite of offices may have to be set up to allow evaluators to complete their tender evaluations in private, so as to maintain the robustness and confidentiality of the evaluation process.

An evaluation model is created using the questions and criteria developed as part of the standard procurement templates and the project-specific requirements. Individual subject-matter experts are then assigned to evaluate individual questions or topic areas in the suppliers' responses. Each evaluator is presented with electronic answer sheets, which contain only the relevant tender questions allocated to them, the suppliers' answers, and access to all of the relevant supplier's uploaded documents. An example of an online eEvaluation page is illustrated in Figure 7.5.

The system controls which evaluators can see which technical questions, and limits their access to only the relevant parts of suppliers'

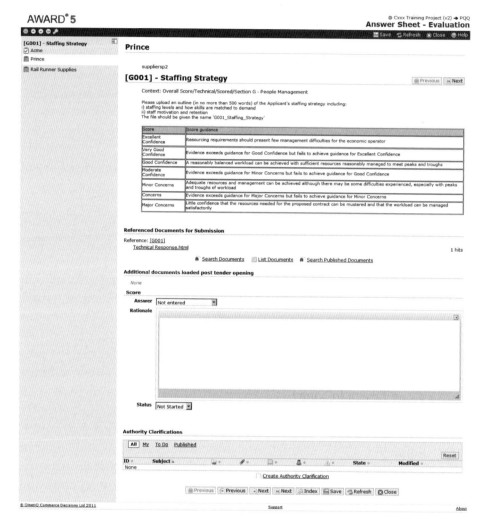

Figure 7.5 Screen shot courtesy of QinetiQ Commerce Decisions Ltd showing their online eEvaluation system, *'Award'*.

submissions. This enables the clients' procurement team to have complete control over the evaluation process.

Like the eSourcing system, the security of the eEvaluation tool is extremely important. Access must be restricted to only those who are evaluating the suppliers' submissions, and evaluators must not be able to view how other evaluators have scored the suppliers' responses or their supporting rationale to those scores. This ensures that the evaluation process reflects the evaluation of each of the subject-matter experts and that it also captures their individual objective rationales. The tool shows how many of the questions have been opened and reviewed, and whether or not the evaluation has been completed. This allows the evaluation management team to monitor and manage the progress of the evaluation in real time, alerting the team of any issues that could affect the procurement timetable, and gives them an opportunity to intervene to ensure timetables are met.

Aspects of managing systematic procurement processes

Standard processes

Standardised procurement processes in a programme tend to lead to consistency in delivery and a reduction in errors. Standardisation gives all stakeholders involved a clear understanding of their inputs and responsibilities. Standardising the processes also gives participating firms clarity regarding the specific expectations and requirements of others. The different steps in the formal procurement process, their associated documents, and an assurance review together form a set of standard procedures for each stage of procurement.

Mobilising large teams of people in a relatively short space of time requires special measures, especially when the people involved are working on diverse projects. This is further complicated by the fact that the staff working on behalf of the client may be either internal to the organisation (directly employed) or external consultants (indirectly contracted and therefore employed by any of a number of outside firms). To reinforce standard procedures, all parties involved need to be issued with a clear definition of the formal practices expected of them. This may take the form of a hard-copy document stating the responsibilities of the various client teams, such as the technical, commercial and legal teams and external advisors, including designers and engineers. Similarly, codifying the procurement procedures has a number of advantages for the client's team. Developing a code of practice formalises the governance and assurance procedures, so that all parts of the client organisation can understand the required steps to be taken to complete a procurement and award a contract. The code of practice can also be used to indicate the skills and experience that are needed by individual members of the procurement team.

Schools of excellence

Managing a programme of procurements makes it possible to aggregate or combine the workload of a series of projects. This gives the client's team the opportunity to develop a schools of excellence model to gather expertise and experience and share it across a programme. Schools of Excellence form discrete entities within the wider body of staff, working on the full procurement process. They operate in effect as a supporting, back-office function, with the rest of the procurement team operating in a client-facing role. The number of schools of excellence within an overall procurement team is determined by the volume and complexity of the procurements to be undertaken. If there is sufficient volume to justify the excellence, then it should be put in place with the aim of helping to deliver robustness, consistency and efficiency.

This is important, as the programme procurement team's primary function is to discharge best practice to meet programme requirements. That involves carrying out the same best-practice activity on a repeat basis, capturing lessons learned through familiarity and delivering consistency in the production of standard PQQs, ITTs, evaluation activities and contract awards.

With the help of a schools of excellence approach, the organisation itself guides its experts or teams of experts, together with appropriate junior staff, to deliver not only best practice but also continuous improvement and increasing efficiency. Using a schools of excellence model both informs the procurement procedure and also helps to define the organisational structure schools of excellence of the procurement function. This schools of excellence model is part of the PSE approach used to develop programme procurement teams.

A possible structure of various schools of excellence is given in Figure 7.4, showing the schools on the left and their functions on the right. One example of a schools of excellence is the evaluation management team. In public procurement, as noted earlier, the risk of potential challenge to the procurement process is always present. The cost of failed procurements in terms of time, money and resources is such that the use of specialist staff to manage the setting-up of evaluation criteria, to review questions and guidance, and finally to assist with the evaluation process to anticipate and reduce the risk of challenge, has been shown to be very cost-effective. Partly as a result of adopting the schools of excellence approach, for the multi-billion pounds of procurement delivered by PSE on London 2012 and other programmes within the

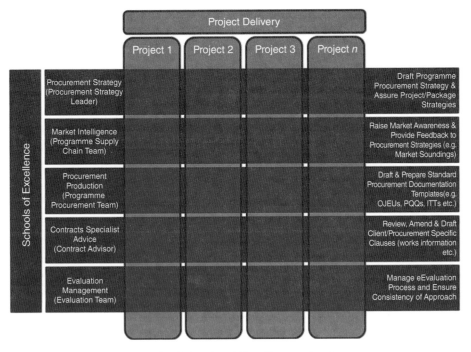

Figure 7.6   An example of a Schools of Excellence structure.

public-sector procurement regulatory framework, none has met with any official challenge at the time of writing. This is especially important on large programmes, where the impact of a challenge will not only delay the individual project in question, but is likely to have a knock-on impact on other projects in the programme and so multiply the negative effects of an individual challenge.

Governance

The checks and balances built into the procurement process are largely determined by the regulatory framework and necessary legal and third-party scrutiny. These checks and balances take the form of standing orders and delegations within the client organisation. The formal procurement process therefore needs to be underpinned by a robust process, which requires that each step in the process meets the obligations of good governance.

The level of scrutiny required for sign-off depends on the price of a particular procurement and the level of risk accepted by the client organisation, although other factors may be taken into account. Individual budget holders are often able to sign off procurements up to approximately £100,000. However, contracts valued at tens of millions of pounds invariably require board or board sub-committee approval. For the system of governance to work in practical terms there must be sufficient time to allow the client organisation to confirm that the requirements are met and, where internal governance processes require approval, time must be allowed for the scheduled monthly board meetings.

Assurance

The procurement assurance process allows for a final check before releasing procurement documentation into the market. This takes the form of an assurance review, which re-examines the entire procurement document pack, including the questions to be asked in the tendering process and the issues that may arise in relation to particular areas for evaluation. The assurance review is often invaluable in uncovering minor inconsistencies and last-minute changes. It is also used as a final test on the compliance and robustness of documentation, ensuring consistency with overall policies and procedures.

This additional step in the procurement process critically assesses and coordinates the work of many people to ensure that the work within their areas of expertise and in relation to their specific details forms a consistent whole and meets the overall objectives of the client. The real value of the assurance review is that it allows the relevant specialists to confirm that their requirements are adequately covered in the documents. It also permits the assurance staff to test the assumptions and robustness of the description of the scope of projects contained in the documents from an independent and unbiased perspective. The assurance process creates a period of reflection to consider the documentation and confirm that the procurement documents are released in a fully formed and well-considered state.

The assurance process inevitably involves a number of people. However, one person should be appointed to act as the assurer for each individual project or package being procured. The assurer then appoints others as necessary to review and test the documentation. By its very nature the assurance process is iterative; the process is repeated until a satisfactory conclusion is reached. The first assurance review may reveal a number of issues, to which the procurement team and the technical experts need to respond and find a resolution. This process itself can take a number of days, editing documents or reaching agreement on issues raised. A second review should yield significantly fewer issues, but could again lead to a further editing of the formal procurement documents.

Although there may be pressure to rush or even cut the assurance process because of scheduled deadlines, documents that are rushed out without appropriate assurance can contain avoidable errors that might have a significant impact on the delivery of the contract. Therefore, a full assurance review needs to be given adequate time as part of the procurement plan, and it also needs to be included in the overall procurement schedule. This final stage of the process before publishing procurement documents is invariably worthwhile.

Training

To take advantage of the benefits of utilising eSourcing and eEvaluation tools, all the processes around procurement need to be standardised. PSE achieves this through the codification of every aspect of the procurement process. All procurement staff, and indeed the whole organisation, need to understand how the procurement process operates and this involves staff training. All coded processes and procedures are based on the requirements and architecture of the eSourcing and eEvaluation systems. In order to be compliant, suppliers are therefore obliged to use the electronic tools available. This drives the adoption of the tools throughout the organisation, which in turn delivers increased efficiency.

The successful implementation of systems such as eSourcing can be achieved only through appropriate staff training. Training needs to provide staff with a good understanding of the system and give users the confidence to operate the system and deliver the required outcomes in the timeframe necessary. The training required should not, however, be overly detailed or onerous; it should cover only the users' specific needs.

The exception to this approach to training is for those individuals tasked with managing procurements through the eSourcing system. Those individuals require more in-depth training – but even their training should still be accomplished in a half-day training course on the basics of using the system. Other staff involved in evaluation should be able to become proficient within a shorter time, and eEvaluation alone can often be covered in no more than one hour, with the briefing taking place immediately prior to their first evaluation exercise, if necessary. The training session should cover not only the use of the software, but also the principles of evaluation. In addition to this a number of staff within the procurement team may need to attend an advanced training course in the use of the system, to enable them to act as in-house experts.

The milestones of procurement reporting

On a large construction programme reporting activities are usually undertaken by a separate functional team called project controls. Project controls essentially gather and present data on time and cost and report on the progress of construction delivery. To enable the consistent reporting of procurement activity it is important to develop and agree a standard set of milestones. The milestones may include (in chronological order):

- publication of a contract notice
- PQQ return
- issue of the PQQ evaluation report
- Tender list approval
- Tender issue
- Tender return, and
- Contract award.

An agreed standard set of procurement milestones can then be embedded within a baseline schedule. The baseline schedule states the expected timing of a series of events and is compared to the actual outcome in a report called the tender event schedule (TES). The TES can be produced every month to organise and update appropriate procurement reporting. The tender event schedule acts as a focal point for other reports, including Board Reports that present project-by-project status reviews against each procurement milestone. These reporting processes are an essential means of giving the programme team an understanding of the key details, derived from the large amounts of information that flow around any large programme. It is therefore critical that these reporting processes are set out early in the programme's life cycle.

London 2012 – Case study

To facilitate the staging of the London 2012 Olympic Games, more than 250 major contracts relating to the Olympic stadia and infrastructure and over 2000 minor contracts were procured in a relatively short period of time. Moreover, all the procurements had to be carried out under public and regulatory scrutiny, as well as under the spotlight of the media and the newly emerging social media sources. The procurement processes had to be undertaken not only efficiently and speedily, but also tactfully, transparently and in a manner that was fair and beyond reproach. The volume of procurements to be undertaken meant that standardisation was the approach used to deal with the documentation, contracts, evaluation criteria and methods of evaluation. Standardisation provided consistency and accountability.

The procurement guiding principles outlined at the start of this chapter were used on the 2012 Olympic programme. The eSourcing and eEvaluation tools were mandated for all procurements undertaken as part of the programme. A procurement code was developed and mandated to ensure that process standardisation and consistency were achieved. The process for procuring London 2012 had to satisfy a significant number of stakeholders with wide-ranging and diverse interests, which were in some instances contradictory.

As a public-sector client, procurement activity was subject to the Public Contract Regulations (2006) relevant at the time. In addition the programme was required to meet expectations regarding innovation and best practice in all London 2012 Olympic construction activity. The requirement to be a best-practice organisation meant that a multitude of bodies and agencies had an interest in influencing the way processes were to be carried out, and as such they all required assurance that the processes adopted were indeed delivering what had been promised.

Standardisation and codification of the procurement process

A procurement code ('The Code') was developed to prescribe how the Olympic Delivery Authority (ODA) was to purchase all works, services and goods. It was based on similar procedures and working instructions from within other safety-critical sectors, including the rail and petrochemical industries. In effect it was an instruction manual embodying best-practice procurement procedures and was based on a 13-step process, with each step having to be signed off before moving on to the next. The steps were developed to ensure strict compliance with regulations and central government requirements in the form of the Office of Government Commerce (OGC) Gateway Process.

The 13 steps in the procurement process were:

- Procedure 1: Establish the procurement route
- Procedure 2: Set contract strategy and criteria
- Procedure 3: Prepare contract notice
- Procedure 4: Prepare pre-qualification questionnaire (PQQ) and invitation to tender (ITT)
- Procedure 5: Issue contract notice and PQQ
- Procedure 6: Receive and evaluate PQQ
- Procedure 7: Agree tender list
- Procedure 8: Notify suppliers
- Procedure 9: Issue ITT
- Procedure 10: Receive and evaluate tenders
- Procedure 11: Prepare tender report
- Procedure 12: Notify tenderers
- Procedure 13: Award contract.

These procedures together provided governance, compliance and auditability, and each required approval. The process also included a set of stage gates at certain points in the process, at which both the ODA and the delivery partner (DP) were required to sign off information before it could be published. The procurement therefore could not progress without their combined agreement at each of the stage gates.

Stage gate 1 – Agreeing the procurement strategy

This stage gate was at the end of Procedure 2, when the ODA's delivery partner (DP) agreed that the procurement strategy was fit for purpose and recommended it to the ODA, who then signed off that they were prepared to proceed on that basis.

Stage gate 2 – Tender documentation completion

The second stage gate, at the end of Procedure 4, followed the preparation of the PQQ and the ITT. After the DP signed off the tender documentation, with particular focus on the evaluation criteria, the

ODA had to agree by stating they were prepared to proceed on the basis of the documents presented. The ITT sign-off almost always took place at a separate and later point in the process than the sign-off of the PQQ.

Stage gate 3 – Tender list agreement

The third stage gate occurred at the end of Procedure 7, to agree the tender list. At this point the DP presented the outcome of the evaluation of the PQQ and made recommendations to the ODA regarding the firms to be included in the ITT list. The ODA then signed off to confirm that the PQQ evaluation process was compliant and that they were prepared to proceed.

Stage gate 4 – The tender report

The final stage gateway to the process occurred at the end of Procedure 11, when the DP prepared a tender report setting out recommendations for the awarding of contracts. The ODA then signed off acceptance of the recommendation on the basis that it had been evaluated in a compliant manner and that the contract to be executed fitted the need identified within the business case, as set out in the project initiation document (PID). Following this gateway the ODA awarded contracts, completing the procurement process and entering the contract administration phase.

Figure 7.7 shows the 13-step procurement process, the stage gateways and the relationship between the ODA and the DP (CLM). The key swim lane is the one labelled 'CLM Procurement', which shows all 13 steps of the process. The stage gateways are clearly shown as the diamonds occurring above procedures 2, 4, 7 and 11. At each of these points the approvals process for both delivery partner and the ODA are clearly mapped. The darts along the base of the process chart summarise the activities taking place at the different stages in the procurement process, from the publication of the contract notice via OJEU to the awarding of contracts. The top swim lane of the diagram shows the functions of the ODA and the DP that were external to procurement, but supported the overall process. The other swim lanes represent the governance responsibilities of the other external functions of the ODA and the DP. The bottom swim lane shows the specific input of the schools of excellence to each of the procedures.

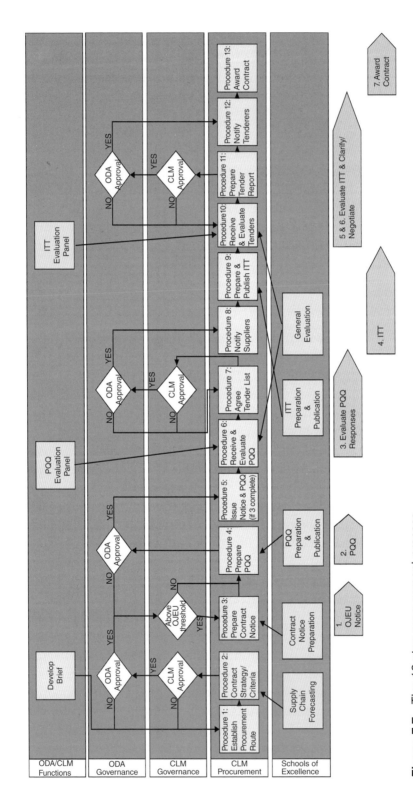

Figure 7.7 The 13-step procurement process.

The code set out specific roles and responsibilities of the procurement staff during the process, ensuring that both parties (the ODA and the DP) were each accountable to the other, while maintaining a fully integrated approach to the delivery of the programme. The use of the code meant that procurement staff in the DP and the ODA were fully aware of the minimum acceptable level of performance they had to meet. Indeed, the code formed part of their job descriptions.

Overall responsibility for procurement rested with the ODA's Head of Procurement, to whom the delivery partner's Head of Procurement and Contracts reported. The purpose of the code was to make responsibilities during the buying process completely transparent and unambiguous, demonstrating clear reporting hierarchies and lines for communication. As the code encouraged all staff to work together in order to generate the information required to satisfy the gateways within the process, it helped to establish an integrated team of the staff at both the DP and the ODA, who were ultimately responsible for the projects. The emphasis on team integration tended to encourage a more cooperative working environment and a sense of ownership within the project teams than might have been achieved in a more formalised client and supplier relationship.

Procurement reporting

Procurement Board meetings attended by both the ODA and the DP were held on a monthly basis. They formed part of a wider programme reporting cycle, which also included a Design and Town Planning Board. The format of the Programme Board was similar to a Project and programme review clinic, where each project reports on progress and issues requiring resolution. The monthly reporting process was supplemented by a weekly high-level procurement update to capture achievements, issues and compliance with plans.

Concluding remarks

The procurement function within a PSE delivery model is prescriptive and requires resources to be invested in its development and codification early in a programme's life cycle. The robustness of a procurement process determines the degree of transparency that can be achieved for audit purposes as well as compliance, demonstrating that procurements are being made in a fair manner.

It is the strategic and procedural elements that make up the purchase engineering part of the PSE model. It should also be apparent that PSE shapes the client's strategy, packaging, contracting, vendor engagement and procurement procedures in order to deliver the client's objectives. Traditional approaches to procurement of projects, portfolios and programmes tend to be based more on procedure and less on strategy, and even those procedures tend to be less prescriptive than those required by PSE.

The use of standardisation and codification is aimed at reducing ambiguity and inconsistency in the procurement process. Consequently, challenges are less likely to materialise. Ambiguity, inconsistency and challenges are clearly damaging to the delivery of the procurement function, especially when the purpose is to attract confidence from suppliers and deliver well-tested, value-based decisions that award contracts to the supply chain that demonstrates the best-value proposition for delivering a client's requirements. The remainder of this book describes how the level of robustness and due diligence delivered in tier 1 procurement may be achieved during the procurement and management of the tier 2 and 3 critical supply chain elements. The following chapter deals with the management of relations with suppliers and how the contribution of each supply chain can be effectively monitored.

eTendering/eEvaluation	
Dos	**Don'ts**
Consider the quantity/complexity of the purchasing requirement when preparing the specification for the system(s) and standard electronic procurement documentation	Ignore the existence of the client's existing systems and check for compatibility/ease of adoption
Prepare a Business Case for using the eProcurement systems for client approval	Underestimate the support arrangements required when developing resourcing profiles for the procurement function
Consider the type of goods, works and services to be purchased and the marketplace when deciding the type of eProcurement tool to implement, e.g. is reverse auctioning likely to be required?	Forget to consider the marketplace and its ability to use an e-tendering system, e.g. are suppliers mainly SMEs without access to IT?
Develop a suite of standard 'non-editable' procurement documents to instruct suppliers on the use of the system and the procurement process	
Limit access for publishing procurement documentation	
Centrally control the development and update/ upload of standard procurement documentation	
Consider reporting requirements pre- and post-contract, e.g. the ability to agree performance measurement criteria at pre-contract stage for eReporting post contract	
Ensure comprehensive training for all users.	
Make sure that whatever e-sourcing system is used is free for suppliers to use and they have access to free support in the use of the system	
Carefully consider what is expected from suppliers and then make sure the questions clearly define the file names of the responses required, e.g. 'Please respond with a file named B01_experience'	
Think carefully about mandatory questions. Ensure that you mandate only what is necessary	Make everything mandatory
Take time to agree with all stakeholders the evaluation model and then test and prove it before live application	

Process Codification	
Dos	**Don'ts**
Review client's existing systems/codes for compatibility, applicability and possible adoption	
Ensure code accords with client governance and financial delegated authorities	
Consider the quantity/complexity of the purchasing requirement when agreeing the degree of codification required	
Codify the pre- and post-contract processes, e.g. procurement code and contract administration code	
Ensure codification fits with the client's quality assurance processes	
Control the development of the code by using a central resource responsible for agreeing and implementing amendments	
Keep the code simple through the use of process flow/swim lane diagrams	
Allow plenty of time to prepare code and to go through a number of iterations before it is agreed by all parties	
Consider developing a 'supplier guide' to inform suppliers on the key procurement processes	
Consider preparing a simple 'buyers' guide' for circulation to 'non-procurement' staff to explain procurement process and provide guidance on smaller value procurements	

References

Public Contracts Regulations, (2006) Statutory Instruments, 2006 No. 5: Public Procurement, England, Wales and Northern Ireland.

Quinn, D., (2012) 'Driving best-practice procurement processes with eSourcing tools', London 2012 Games construction project learning legacy paper. http://www.legislation.gov.uk/uksi/2006/5/pdfs/uksi_20060005_en.pdf (accessed 1 September 2012).

8

Managing supply chain involvement across a programme

The Queen's Diamond Jubilee Royal barge, The Gloriana, was moored in the Olympic Park during the Games (photo courtesy of AECOM).

Programme Procurement in Construction: Learning from London 2012, First Edition.
John M. Mead and Stephen Gruneberg.
© 2013 John Wiley & Sons, Ltd. Published 2013 by John Wiley & Sons, Ltd.

More than 2500 sections of steelwork were installed to complete the steel structure of the Velodrome (photo courtesy of Mark Lythaby).

Introduction

In earlier chapters the communication of a client's values was described and how these values were used to influence the packaging of the programme for procurement. The book has also discussed how contract strategy is formulated to deliver these client value requirements. The activities described in Chapter 6 on supplier engagement identify those supply chains with an interest in the programme opportunities, and subsequently how the procurement process and procedures are deployed and managed using eSourcing.

How appetite is generated during the procurement planning phase and how the procurement plans for individual procurements are tested through the use of the market soundings exercise have also been described. The market sounding exercises help to anticipate which organisations may be interested in responding to the procurement once it is published. However, once a tender opportunity is published and is live in the marketplace, no further dialogue around the procurement is possible that might offer an advantage to one bidder over another and lead to a challenge of the award decision later.

View from London Way back down to the Olympic Stadium (photo courtesy of AECOM).

Of course, that is not as great an issue for the private sector, which falls outside the jurisdiction of the Public Contracts Regulations and is therefore much less regulated. Private-sector procurers have much more flexibility to negotiate the price and play one bidder off against another. However, while this kind of commercially led negotiation may reduce the price of a tenderer's offer, the outturn cost is often greater than the tendered figure, either through increased claims or a reduction in the quality, or even the non-fulfilment of other value-adding elements. This chapter describes how the appetite of supply chains is monitored and managed as the programme procurements progress.

Supplier relationship management

Once a procurement reaches the stage of inviting tenders, the process of bidding, particularly when using eSourcing systems, can become impersonal. That is because the procurement process must be compliant, secure and above all fair to all participants; also, any conflicts of interest need to be identified and removed from the procurement process in order to avoid possible issues or the risk of legal challenge. As an added level of probity, no direct communication with tenderers can be made outside the official electronic tendering system. This impersonal approach can lead tenderers to feeling cut off from the client and the procurement team to some extent.

However, the programme supply chain team are in a useful position at this point in the procurement process, as they have worked on engaging many of the possible suppliers, and can offer an indirect route of contact to the client and the procurement team. The programme supply chain team can also act as a sounding board and guide. Because of the relationships formed with the programme supply chain team during the vendor engagement phase of the programme, the supply chain team are therefore often contacted during procurement by bidders seeking advice. Nevertheless, the rules of engagement during this time are strict and the programme supply chain team cannot offer any advice directly related to any active procurement.

Although their role is restricted, they can direct bidders to the correct channels and to relevant sources of publicly available material, including the supplier guide or the client's balanced scorecard. In practice, the programme supply chain team will always direct bidders to post a request for clarification on the electronic tendering portal or via other official and auditable routes. Where the query is not directly related to the procurement, then they can assist in keeping suppliers engaged by understanding their intent in relation to the submission of a PQQ or a response to an ITT. This intelligence is vital during procurement, because until the deadline for responses is reached, the number and identities of bidders may be unknown. Having an understanding of whether those suppliers that have expressed interest to date will actually submit a tender as the deadline approaches is vital to avoid surprises.

During procurement the supplier relationship developed during the vendor engagement phase of PSE can be used to keep bidders on track and engaged. However, it is important that communication lines are fully controlled, and therefore the two key routes used are through the programme supply chain function and the electronic tendering portal. The clear rules of engagement and communication mean that the trust of suppliers can be developed and reinforced, both in the process and more importantly in the client as a best-practice procurer, safe in the knowledge that a fair, transparent and auditable procurement process is being delivered. After a contract has been awarded, this trust between suppliers and the process allows for robust post-tender feedback and, in public procurement at least, it contributes to the confidence of tenderers that the results are fair, which reduces the risk of a challenge from a losing organisation. A challenge to any procurement within a programme can add delay, cost and uncertainty to that programme, all of which are risks PSE seeks to avoid.

Remaining in contact with all firms who tender for work

To meet the aim of the purchasing function of PSE to be equitable, transparent and fair to all organisations expressing interest and submitting tenders, one minor problem needs to be overcome. Once any procurement is live, it must remain equitable and all communication between tenderers and the procurement team must be transparent, auditable and equal. This is necessary to give the firms on the supply side confidence that they are competing in a fair competition, in which they all have an equal chance of winning.

To comply with EU legislation, a tender opportunity that is above the threshold limits must be published in a contract notice via the Official Journal of the European Union (OJEU). The contract notice, which is drafted by the programme procurement team, contains details of the opportunity and a link to its eSourcing portal. Once the tendering procedure begins, a list of those organisations that have already registered to view the opportunity via the eSourcing system can be reviewed by the programme supply chain team and compared with the list of organisations that expressed interest during vendor engagement. Any firms that showed interest during supplier engagement, or participated in the relevant market sounding exercise but do not appear to have registered on the eSourcing system in response to the contract notice, can be contacted and made aware of the opportunity if necessary. This serves either to prompt them to register, or at least to ascertain why they might not have done so already. This can be a particularly delicate stage in procurement and may lead to a further reassessment of market interest in the opportunity to understand any change in the appetite of firms. This early intervention can identify any organisations that may have changed their stance on the business opportunity offered since the market sounding was taken. The procurement team can then estimate the likely number of organisations that may respond to the tender, thus allowing time to address any issues that might emerge through a lack of competition in the event that the appetite identified earlier is not forthcoming.

The point of maintaining a close overview of the procurement process at this early stage is to remove any surprises, if at all possible. Clients need to stimulate a sufficient response to their tender opportunities to create a competitive tension that will increase the value for money of the wining bid. It is therefore important that the programme knows of any organisations that may have indicated an interest in tendering but have not registered on the eSourcing tool, or might not have viewed the associated documents, while there is still time to encourage participation. At the very least the client can gain an understanding of why appetite might have changed.

Tracking bidders in this way and gaining feedback from organisations that may have indicated a positive appetite during the market sounding exercise, but that have since decided not to respond, could also assist with future procurements. The information can indicate how successful or not the procurement team were in responding to the feedback obtained from market soundings and could even offer an insight into the current state of the market beyond the programme. Appetite is a function of client demand and, if the general demand for construction works is low, then supply and the appetite to compete for contracts is likely to be relatively high. Equally when demand for construction is high, appetite is likely to decline as the capacity of supplying organisations to take on additional work is eroded by full order books. This could lead to suppliers to target carefully those opportunities that they deem to be the most attractive in terms of winning and of generating the greatest returns.

Case study – London 2012 Olympic Delivery Authority

The effect of the construction market was very evident in 2006 and 2007, when the Olympic Delivery Authority (ODA) began to engage the construction sector to procure the venues and infrastructure for the London 2012 Olympic and Paralympic Games. The ODA used a programme approach to procure and manage the capacity and capability of the supply chains during the games procurement and construction, as it was important that this particular public-sector-funded programme was delivered on time and to budget, while also delivering on the promises made in the original bid to host the Games in terms of legacy, regeneration and sustainability.

The challenge of attracting suppliers to bid for the ODA's opportunities was that the construction sector was experiencing a peak in demand at the time. There were many private-sector opportunities for the supply side to pursue that offered less bureaucratic intervention, as private-sector projects are not subject to as many regulations. More importantly, they offered the same or less risk and greater potential for reward. The ODA at the time was perceived as a one-off client, which was procuring a number of large and complex construction and civil-engineering projects on a highly contaminated site in a densely populated area of east London. Many of the opportunities on offer were sports stadia, and the well-publicised failure of the Wembley Stadium project to realise its cost and schedule targets was still fresh in the minds of those operating in the sector. Add to those factors the absolutely immoveable deadline for delivering the Games, and it becomes clear why appetite for ODA projects in 2006 was low.

Through a strong market engagement plan, one by one the perceived fears of the market were addressed through vendor engagement techniques, including industry days, Meet the Buyer events and an open two-way communication with the market. These methods did not convince all organisations that the ODA's demand was appropriate for them and, while many firms were engaged at that point, only some expressed interest and actually went on to respond to opportunities when they were published. However, throughout this period the procurement team were always aware of which organisations were interested in participating and which were not. The procurement team also knew the reasons behind their decisions up to the point of the tender receipt deadline. Bidder tracking had succeeded in avoiding surprises and, where appetite was low, decisions were taken to adjust procurement plans to raise appetite to attract bidders to respond and ensure competitive returns.

Through the transparency and robustness of the interaction with the supply side, the ODA was able to demonstrate value for money against its balanced scorecard of value criteria. It was able to procure the most capable and appropriate organisations to deliver the buildings and structures as specified, ahead of time and in many cases below budget. It was also able to procure over 150 major construction contracts, worth many billions of pounds, without a single challenge.

According to the National Audit Office (2008), reviewing the ODA's approach to procurement, there was clear evidence that procurement and supply chain management had evolved to an advanced level 'with innovative processes in place'.

Supply chain mapping

On any large programme composed of a number of different projects, it is of paramount importance that a strategic overview is maintained of who is bidding for what and who has won which contract. This overview of directly procured contracts is achieved at the tier 1 level by mapping the supply chain during procurement up to the actual award. This map is a single source of information that changes gradually over time. The tier 1 supply chain map also feeds the data that is used to compile and publish the opportunity slides that were produced as part of the vendor engagement.

Supply chain maps make use of the packaging strategy developed as part of the overall procurement strategy. These supply chain maps reflect the clusters or categories of projects and packages of work as they are grouped together, either according to different types of work or a number of other criteria, such as geographical location. A programme clustering model may be comprised of one or more of the following cluster headings, which group together individual contract opportunities. Examples of these clusters include:

- Design contracts, where the level and kind of design packages depend on the level of design to be procured and the procurement routes being used;
- Enabling works contracts, including items such as demolition and remediation;
- Civil engineering contracts, which may cover packages such as roads, bridges and tunnels;
- Construction contracts for buildings, accommodation and other vertical build facilities;
- Utilities contracts, including utility networks and associated infrastructure;
- Common components and commodities, where these may be, for example, specialised items with a long lead time, against which leveraging greater demand could prove of benefit to the overall schedule of the programme. Specific examples include specialist M&E components or simple standardised commodities that can be bought strategically across the programme in high quantities, whose standardisation will benefit supply and can be specified simply – for example, materials like aggregates or steel reinforcement;

- Logistics contracts, which apply to items that are required across the programme by all or many contracts and may include some preliminary-type packages, such as site accommodation, security items, welfare facilities and transportation.

Figure 8.1 shows an example of how one such cluster for a tier 1 supply chain might be mapped, and Figure 8.2 shows a screenshot of a full tier 1 map for the Crossrail programme, taken during the spring of 2012.

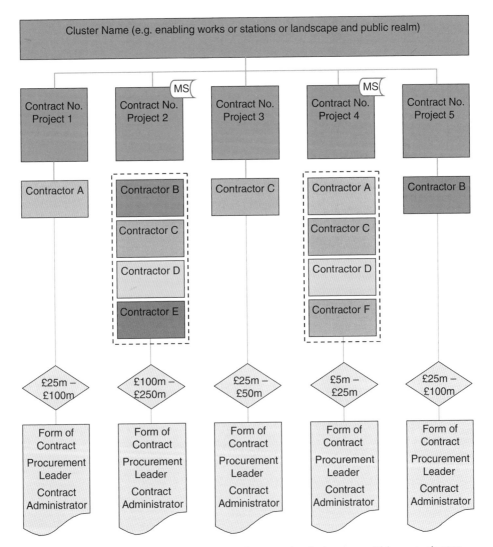

Figure 8.1 Example of a tier 1 contractor supply chain map within one cluster category.

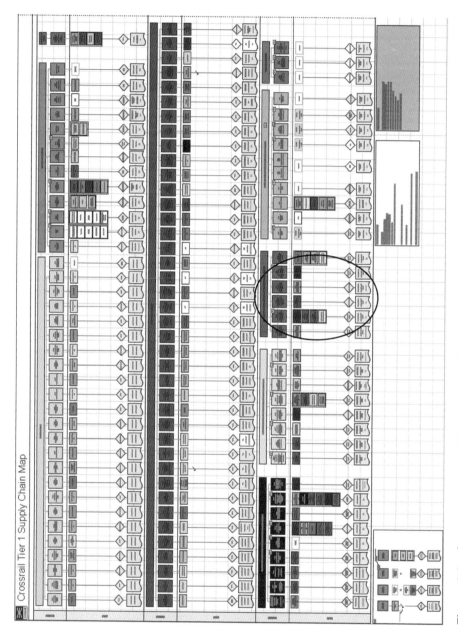

Figure 8.2 Screenshot of a full tier 1 supply chain map for the Crossrail programme.

Figure 8.1 shows a cluster of five related procurement packages including five contracts, with three of those contracts awarded. Only one contractor appears beneath each awarded contract. Project 1 is awarded to Contractor A, Project 3 to Contractor C and Project 5 to Contractor B. The map also shows that Project 2 and Project 4 are currently being tendered and the contractors are shown in the boxes below these projects. Each of the contractors' boxes in the diagram is colour-coded for ease of identification. This is especially useful when the full map is viewed and visual identification of where the same contractor appears more than once is required. The tier 1 supply chain map also acts as a quick reference to assist in identifying the extent of the appetite of individual firms across the programme and clusters.

The five packages to be procured within the cluster in Figure 8.1 are also all colour–coded, in the same colour as the cluster within which they sit. Each of the packages is given a contract number and a title describing the contract, according to the programme work breakdown structure. Beneath that is a diamond-shaped box showing the value range for the contract and beneath that in turn is a box showing the type and form of contract under which the award has been, or will be, made. This box also includes the name of the procurement leader for the project and the contract administrator for the contract. Capturing these key individuals on the map identifies them within a programme, which may have many hundreds of individuals involved, and again it serves as a quick reference guide for users.

The information presented in the supply chain map is also made available publicly to any interested parties, on the programme website or upon request by email, in a slightly different format as opportunity slides. Publicising information relating to the procurements across the programme, including the names of tendering firms and those that have been awarded contracts, enables the lower tiers of the supply chain to contact the tier 1 contractors and offer to support them by providing key supplier input into their tenders or simply marketing their services. This early engagement of firms from the lower tiers also creates an opportunity for innovation to be introduced at the tender stage. The lower tiers of the supply chain tend to be the smaller and more specialist firms with key product and system innovations, but owing to their size they do not employ large business development teams. As a result these organisations often do not know whom to contact and consequently miss opportunities for themselves to add value during the tender development stage. It is also often difficult for such organisations to gain access to tier 1 organisations so as to understand the potential for a future opportunity at the right time, when product purchase decisions are being taken or when it might be best to make contact with the tier 1 integrator at a later date. This kind of visibility of tier 1 tenderers for tier 2 innovators can improve the tier 1 firm's chances of success and the client's aims for achieving improved value.

Another item to highlight in Figure 8.1 is the call-out flags, which are labelled 'MS'. These appear next to Projects 2 and 4. The flags capture the fact that market sounding exercises were completed during the development of the procurement plan, and they also show that bidder tracking is available to the procurement team during the procurement process. Figure 8.1 is a sample detail taken from a model supply chain map of a programme, showing only one cluster. Figure 8.2 is an illustration of a complete tier 1 supply chain map for a major programme. This is a deliberately illegible screenshot of the whole of the Crossrail tier 1 map. The block circled in the lower half of the map in Figure 8.2 indicates where the cluster detail in Figure 8.1 might fit into the overall scheme of the programme.

The document is actually printed to an A0 poster and displayed at various points in the programme office. For all the executive team and senior programme managers at both the ODA, where this type of tier 1 supply chain map was first developed, and also at Crossrail, the poster was a regular feature in offices and on walls in the programme's office buildings. It served as a simple yet effective overview of procurement progress, areas of overlap between tier 1 contractors and a high-level view of areas of potential risk and opportunity that could be further explored as they emerged on the map. It was also a means of communicating the scale and significance of the programme to all levels in the procurement team. As can be seen from the screenshot in Figure 8.2, the Crossrail programme involved large numbers of procurements and the potential for much supply chain overlap. A silo project approach to such a programme would make the strategic view of the whole very difficult to manage.

This kind of mapping exercise shows the stage of each package in relation to the procurement timetable, including which contracts have been awarded, what is currently being procured and also what has yet to be procured. It also highlights the contracting strategy for each procurement, the approximate value and key personnel (including the procurement leader and contract administrator). However, these are only some of the details that the tier 1 map can capture. The basic model shown here can be expanded to capture other details of each project and package.

Concluding remarks

The PSE approach to vendor engagement and supplier relationship management during procurement allows both trust and influence to develop. The supply side of a procurement want to know they are being listened to. They want to know there is a person who can be contacted during procurement to gain advice that will not jeopardise or break any rules of engagement. Nevertheless, they often need advice that will assist them in understanding what is required of them in a highly regulated procurement environment governed by stringent legislation. The PSE model promotes this through the programme supply chain function – a team that can provide this open channel of communication while ensuring compliance is maintained throughout.

Mapping the programme's emerging tier 1 supply chain in a simple format that conveys the complexities of the programme's packaging and contracting strategies informs the wider programme management team of the 'big picture' and, most importantly, how that picture is evolving. The fact that this big picture is constantly changing means that the tier 1 supply chain map also helps the wider programme team in understanding procurement progress. It also gives vital information about the programme in relation to the market in general and in particular the key suppliers who are engaging in the programme. The project teams can see how they fit into the overall grand scheme of the programme, and the map can also highlight the relative importance of potential impacts and consequences of each package on the overall programme's future success or failure.

Appetite Management	
Dos	**Don'ts**
Provide a point of contact outside the procurement process, e.g. independent of the Procurement Manager	Be afraid to ask any organisation whether they are or may be interested, even those that seem incompatible
Act as a conduit for any significant strategic matters that may arise	Jeopardise the procurement process by your actions, if unsure on the correct procedure, ask before taking action
Report appetite intelligence to procurement manager throughout the procurement process	
Model impact of award across all vendors and all areas of appetite	
Treat all prospective suppliers equitably	
Allow meetings to be informal in order to promote open and candid dialogue, foster confidence and maximise their usefulness	

Reference

National Audit Office, (2008) *Preparations for the London 2012 Olympic and Paralympic Games: Progress Report*, June, London, NAO Report.

9

Due diligence and the management of capacity

View of the Olympic Stadium from inside the Aquatics Centre roof structure (photo courtesy of Mark Lythaby).

Programme Procurement in Construction: Learning from London 2012, First Edition.
John M. Mead and Stephen Gruneberg.
© 2013 John Wiley & Sons, Ltd. Published 2013 by John Wiley & Sons, Ltd.

View of the diving pool from inside the Aquatics Centre structural steel roof structure (photo courtesy of Mark Lythaby).

Introduction

The previous chapter discussed the monitoring and management of relations with the supply chains in a programme with the use of supply chain maps and other tools. However, during procurement the capacity of tier 1 contractors to deliver is also subject to testing, as part of the due diligence required by public-sector clients. This chapter describes how this aspect of due diligence is modelled and this leads on to a discussion of risk and exposure as it is applied to the critical supply chain elements at tier 2 and tier 3, both during their procurement by the tier 1 contractors and also during delivery. This monitoring of the supply chain as it is assembled by the tier 1 contractors assists the firms involved in projects and packages to understand their place in the programme and the impact they may have on its relative success or failure. It also enables the programme management team to work with the projects to avoid or mitigate emerging supply chain risks across the programme.

Modelling supplier utilisation

Tier 1 supply chain maps work well visually. They are quick and simple reference documents for large construction programmes that highlight

Aquatics Centre main pool almost completed and ready for the water – in total more than 180,000 tiles were used in the pools (photo courtesy of Mark Lythaby).

areas of overlap across contracts by similar contractors and suppliers. However, the precise extent of the exposure of tier 1 contractors across the programme remains unclear. To overcome this weakness, behind the poster image of emerging contracts of the tier 1 supply chain map sits a supplier utilisation model.

The supplier utilisation model compares the capacity of each tier 1 contractor with their actual exposure (awarded contracts) and their potential exposure (the contracts being tendered by the same firm across the whole programme). It effectively translates the picture that is drawn for the tier 1 supply chain map into a measure of the programme exposure of each tier 1 contractor. This highlights the potential risk of over-utilisation of any one individual contractor across the programme at any one time.

A utilisation model uses the ratio of the budgeted annual cost of projects or packages of work that a supplier has been awarded, or for which they are currently tendering, over the supplier's latest published or average annual turnover. Turnover is used as a historical indicator of a supplier's capacity. It is a factual source of the past amount of work completed and delivered in the previous year, and demonstrates that the supplier had at least sufficient resources to complete a certain level of output. While turnover in the past does not guarantee future output, it does give a good indication of the level of output suppliers are capable of delivering.

There are three ratios, and which one is used depends on the question posed, the timing and the circumstances. The ratios are:

$$Utilisation_1 = \frac{Actual\ Contracts\ won}{Turnover}$$

$$Utilisation_2 = \frac{Tendered\ Contracts}{Turnover}$$

$$Utilisation_3 = \frac{Actual\ Contracts\ won + Tendered\ Contracts}{Turnover}$$

While PSE aims to ensure that appetite is increased during procurement across the programme for all projects and packages, this has to be balanced against the potential for overexposure of any one supplier to the programme and to any one supplier by the programme. If a client's values are to be delivered, then it is argued that they need to maintain a dominant position in relation to their suppliers. The interdependence of the client and supplier requires an equitable transfer of risk. Exposure or dependence becomes an important part of the risk equation. The aim is to achieve a balanced and mutually healthy interdependence.

According to the OGC (2008), a safe annual contract limit should be no more than 25 per cent of a firm's turnover; otherwise the firm could become 'over-dependent' on the contract or the client. The OGC added that this financial criterion should be considered alongside other questions, including whether or not the firm had the financial strength to carry out its obligations under the contract and whether or not the firm had the capacity to carry out the work.

This guidance reinforces the parameters used by PSE when modelling utilisation, as PSE uses the 25 per cent level of exposure for any one supplier for any one contract as an early warning for over-dependency. This level, however, is often raised to 30 per cent or more in private-sector organisations for individual contracts, and depends on the attitude to risk and other commercial and strategic concerns of the buying organisation.

The 25 and 30 per cent levels of exposure are sufficiently significant to incentivise suppliers to mobilise their best teams to meet client demand, firstly during the tendering process and secondly during delivery. At the same time these levels of exposure are not so great that a supplier's output might be too concentrated in one contract, but they leave them with the opportunity to diversify across other markets and clients.

Figure 9.1 demonstrates the implications of the interdependencies when the forces of supply and demand shift from a buyers' market to a sellers' market, and also highlights how the dominance between the client and the seller affects the power position in the demand and supply relationship.

In the top left quadrant, shown as, 'Demand Dominance', the balance of power lies with the strong buyer, who has potential influence over a weak supplier. This dynamic offers significant opportunities for leveraging greater client value from supply, and that can be achieved through aggregating demand or targeting suppliers whose capacity is appropriately matched to the scale of demand.

This contrasts with the lower-right quadrant, which illustrates 'Demand Dependence'. In this situation the balance of power lies with

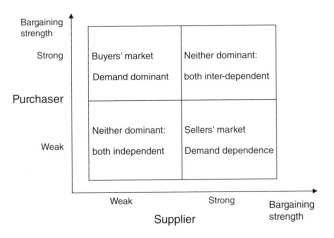

Figure 9.1 The relative bargaining positions of buyers and purchasers.

the strong supplier and therefore it is they who have influence over the buyer. The relative strength of the supplier offers very limited opportunities for the client to leverage greater value from supply and can leave the client exposed to supply restrictions or unanticipated higher costs.

A third situation is shown as 'Demand–Supply Independence' in the lower left of Figure 9.1. In this relationship, neither the weak buyer nor the weak supplier is of sufficient significance to the other to hold any position of power in terms of leverage. The buyer may wish to have greater leverage, but cannot offer sufficient opportunities to the supplier to impose any solution in their own favour. This situation puts both parties in a position to achieve a Nash equilibrium, which leaves both sides optimising their own position given the constraints imposed by the other party. However, it does not always follow that firms choose the best option for themselves. A detailed explanation of the Nash equilibrium is beyond the scope of this book.

Finally, in the upper-right quadrant, labelled, 'Demand–Supply Interdependence', mutual benefit can be gained by both parties as both the supplier and buyer are strong and of enough importance to each other to be able to deliver value. As in the third situation, both sides are again in a position to achieve a Nash equilibrium, which may well be said to occur as both parties are in a position to negotiate.

This model of power relations in supply and demand, when applied to a programme of works, must also consider the power of each project and its relative importance to the whole programme. The degree of power held by a client varies depending on the stage a programme has reached. The client may also be able to increase their power as a result of the way they package their demand. This is the reason the packaging strategy is important when setting out an overall programme procurement strategy. The client may also be able to assert their power during delivery. For example, early in the programme, when several contracts are being processed, the client can exercise power by ensuring the client representatives also communicate the total size of the demand prize to the supply side. There is therefore a need within PSE to ensure the market is fully engaged and understands the full extent of the work on offer.

As the programme progresses and contracts are awarded, it remains important to monitor exposure and supplier utilisation. This provides visibility and an understanding of where the balance of power rests at any one time. In any case, in the public sector, contracts must be awarded to the supplier providing the most economically advantageous tender (MEAT).

This condition may place the buyer and supplier in any one of the quadrants of the matrix in Figure 9.1. Using this theoretical framework to understand power relationships between buyers and sellers is useful for assessing where relative power resides. Moreover, by modelling supplier utilisation the programme supply chain manager can avoid surprises by asking pertinent questions at the appropriate time during procurement rather than during delivery, when it may be too late. Awareness of supplier utilisation allows the proactive management of risk. While it may be possible to avoid a supply risk during procurement, after a contract has been awarded it is possible only to mitigate the impact of that risk.

However, the utilisation ratio of future output should be viewed in the context of the economy as a whole. For example, using this ratio to model exposure during a period of economic growth immediately after a recession could indicate falsely that a supplier was overexposed. This misinterpretation of the ratio could be due to an earlier contraction of the wider economy and a deliberate commercial decision to downsize activities of the firm in line with falling demand. That decision could have been part of a corporate strategy of a well-run company to conserve resources in readiness for future activity. Therefore, in this situation, it may be prudent to review the supplier's turnover prior to the economic contraction, using the turnover of at least the three previous years. The utilisation ratio could then be calculated using the peak year or the average of those three previous years. At the same time questions need to be asked regarding capacity, whether it had been lost or whether it could be remobilised. These are important issues, particularly when the organisation in question provides physical production assets – such as plant and manufacturing equipment – rather than labour, which can be more easily mobilised in a shorter period whenever increases in demand occur.

The following table and diagrams set out a scenario where four contractors respond to five projects in a hypothetical programme with several large projects to be delivered over a period of years. Table 9.1 sets out the data related to their capacity and the demand to which they are responding. The graphs that follow show how the supplier utilisation data might be modelled and interpreted.

Table 9.1 Capacity of contractors viewed against projects

Suppliers and their turnover		Project 1			Project 2			Project 3			Project 4			Project 5		
		Yr1	Yr2	Yr3	Yr1	Yr2	Yr3	Yr1	Yr2	Yr3	Yr1	Yr2	Yr3	Yr1	Yr2	Yr3
Contractor A	300	20	15	10	x	x	x	15	35	5	10	15	5	0	5	20
Contractor B	120	x	x	x	x	x	x	15	35	5	10	15	5	0	5	20
Contractor C	150	x	x	x	x	x	x	15	35	5	x	x	x	x	x	x
Contractor D	100	x	x	x	9	10	2	15	35	5	10	15	5	x	x	x

In Table 9.1 projects (1 to 5) are shown in the columns and the contractors (A to D) are in the rows. The turnover or historical capacity of each contractor is given in the column next to each supplier. The table also highlights which contractors have been awarded which projects. For example, Contractor A has been awarded Project 1. The other firms have therefore been greyed out, as no other suppliers have cost flows allocated to them. The table also shows which suppliers are bidding for which projects. For example, all contractors are shown as engaged in tendering for Project 3.

In Figure 9.2 the vertical axis shows the percentage of the expected annual utilisation of each firm compared to their capacity. In Figure 9.1 it is assumed that each firm is awarded all the contracts for which it is tendering. The graph compares the future predicted cost flow (the expected maximum demand facing each firm) over its duration – in this case three years – with the estimated capacity of the responding suppliers to deliver (the potential supply).

As this model uses the previous year's published turnover as the measure of each firm's capacity, it would require updating annually to reflect the most recent published turnover of each supplier.

The utilisation graph in Figure 9.2 shows each supplier's exposure as a percentage of their turnover for each of the three years. The 25 per cent threshold of over-dependence is also shown. This is indicative of the relative power position and buyer–supplier interdependence. However, the graph does not show the number of contracts included in each firm's exposure; neither does it show the extent to which each supplier's capacity has been secured through awarded contracts and contracts currently being tendered. To see this breakdown of the status of contracts, Figure 9.2 needs to be read in conjunction with Figure 9.3,

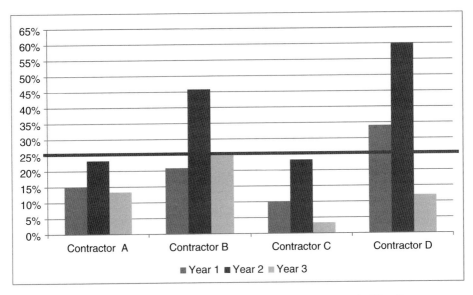

Figure 9.2 Contractor exposure, showing peak exposure and utilisation percentages over time.

which shows an appetite graph. In Figure 9.3 the number of projects is given on the vertical axis. '*T*' projects are those being tendered and '*A*' projects are those that have been awarded.

Figures 9.2 and 9.3 together present the supplier utilisation model, used to assess each contractor and their relative exposure across the programme. Based on Figure 9.3, the report on the above hypothetical programme would show that Contractor A has been awarded one project (Project 1) and is currently tendering a further three projects across the programme (Projects 3, 4 and 5). If they were awarded all of the projects they are currently tendering, on the basis of the information in Table 9.1, their utilisation would not pose a risk of overexposure, as they have the notional capacity to deliver all four projects. The peak programme utilisation for Supplier 1 would occur in year 2, when it reaches between 20 and 25 per cent of their previous year's turnover.

Similarly, Contractor B has yet to be awarded any projects on the programme but, according to Figure 9.3, is currently tendering for three projects (Projects 3, 4 and 5). If they were to win all the projects they are currently tendering, the peak utilisation would occur in year 2, when it would be expected to exceed 45 per cent of their most recently published turnover.

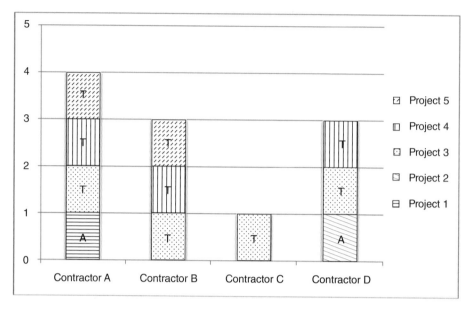

Figure 9.3 Appetite graph, showing awarded and tendered projects against engaged contractors.

Meanwhile, Contractor C is currently tendering for only one project on the programme (Project 3), which if they were to be successful would represent an expected peak utilisation in year 2 of between 20 and 25 per cent of the firm's capacity.

Finally, the model indicates that Contractor D has been awarded one project (Project 2) and is currently tendering for a further two projects across the programme (Projects 3 and 4). If they were awarded both of the projects they are currently tendering, in addition to the contract awarded for Project 2, their utilisation could pose a potential risk of overexposure. The expected peak programme utilisation for Contractor D would occur in year 2, when it reaches 60 per cent of their capacity as indicated by the company's previous year's turnover.

This analysis equips the programme team with the information needed to ask relevant questions of their suppliers during procurement and to also assess the relative power of the various suppliers during delivery and help to prepare for any associated negotiations. The aim at this stage of the programme is to avoid the risk of over-utilisation and a situation in which the buyer is in a weak negotiating position. While this would not be to the buyer's advantage, it could be equally damaging for the supplier.

From this analysis of capacity and utilisation, the programme team are now equipped to develop a number of questions for the suppliers of each of the procurements. For example, as far as Contractor 1 is concerned, they appear to have sufficient capacity to deliver their current contractual obligations and potentially the additional three projects for which they are currently tendering. However, it may be prudent to check which individuals are being proposed to manage the contracts in their tenders for each of the tendered projects. That would reveal whether or not they were over-selling certain individuals' expertise, who may already have been engaged on another awarded project. At this stage it might also be worthwhile to ask the supplier to identify its key supply chain members and how they proposed to manage any overlap across the multiple projects for which they were tendering.

In the case of Contractor B, as Figure 9.3 shows, this programme alone would be drawing on more than 30 per cent of the supplier's resources to deliver its contracts in year 2. It would therefore be reasonable to ask how the supplier intended to meet that amount of demand in one year and how it intended to mitigate the risk of allocating so much of its resources to one client. Again, it may be valid to enquire about the key project individuals allocated to each tender, to ensure that the work could be managed in the event of an award of more than one project.

Contractor C appears to have targeted a specific project on the programme. If a Market Sounding exercise had been carried out for this project, that assumption could be confirmed. Their size and the size of the project seem well matched and therefore this supplier should be able to put together a robust and well-focused response.

Contractor D already has a contract that does not exceed its acceptable utilisation level, according to Figure 9.2. However, if this supplier were to be awarded the other two contracts it is currently tendering, its over-exposure threshold would then be crossed. In anticipation of that possible overexposure, a number of issues call for immediate clarification. For example, the contractor needs to explain how further resources would be mobilised to respond to the increase in demand (assuming they have other clients beyond this particular buyer's programme). Contractor D also needs to convince the programme that they genuinely want to win both projects. Alternatively, the contractor should be asked whether it would be better to focus on one project, which would enable them to remain within an acceptable utilisation threshold.

A programme, portfolio or client that has multiple projects to manage should take a strategic view of their exposure to the supply chain.

During the procurement process, only the utilisation of tier 1 suppliers can be accurately modelled. In themselves construction firms are very rarely vertically integrated, with wholly owned or internal supply chains, and so the capacity demonstrated by a previous year's turnover is in fact actually delivered by many diverse subcontractors and suppliers. Gaining visibility of these subcontracted parts of the delivery team is the next phase of PSE and the programme supply chain team's work. Because there are far more tier 2 and tier 3 firms in the supply chain than tier 1 contractors, the supplier utilisation model can be expanded exponentially, depending on how far down the supply chain the programme supply chain management needs to investigate and have the ability to influence. The actual process of mapping tier 2 and tier 3 firms follows the same procedure as with the tier 1 contractors, although a map is not drawn as it would be of little use. Instead, the utilisation of the tier 2 and 3 supply chain is measured in terms of their utilisation against each project and cumulatively across the programme.

Monitoring the financial strength of suppliers

Monitoring the capacity of a contractor allows problems to be anticipated and action to be taken in good time. The capacity of a firm changes over time and may very well be influenced by factors outside the programme's control. However, an even more volatile measure of a tier 1 supplier's ability to deliver a contract is their financial strength. This aspect of an organisation represents as much risk to the programme as the firm's physical capacity, especially in periods of economic change and uncertainty.

Financial checks of tenderers need to be completed during procurement by the programme procurement function before large and complex contracts can be awarded. The information available to carry them out is usually the last three years of published annual accounts. Third-party credit ratings agencies also use these and other data elements when calculating their ratings. The tests of financial strength and solvency involve calculating a number of conventional accounting ratios, including gearing, solvency, and the acid test, many of which are recommended by government good-practice guides, such as the Office of Government Commerce's *Supplier Financial Appraisal Guidance* (2008).

On large projects the management of risk involving a supplier's financial failure may require some kind of performance bond or parent company guarantee. Crossrail, for example, had a policy of seeking a parent company guarantee from the global ultimate parent of all its tenderers, and also established the health of the global ultimate parent by reviewing and analysing their published accounts from the previous three years. It is debatable whether or not it is strictly necessary to apply such rigid checks on large, global, tier 1 suppliers. However, the volatility of the global economy following the 2008 financial crisis and the failure of organisations deemed 'too big to fail' has made such risk-management procedures increasingly necessary.

One of the problems with analysing and checking a company's historical accounts is that they give historical performance only on an annual basis at worst, or an interim basis at best. The timing of this information for a full year's trading report also depends on the accounting year of the company. Because of the timing of the programmes' own procurement activity, it could mean that the most recent set of accounts provide trading figures that are more than a year old. However, the economic situation in general, and trading conditions facing individual firms in particular, can change in the period of a year. For example, during the 'credit crunch' financial crisis beginning in 2007 and 2008, as its name suggests, access to credit facilities to finance production activity became difficult to obtain. As a result, organisations looking to meet their cash-flow requirements prior to being paid by customers could not rely on being able to borrow from the banking system.

To monitor financial aspects of supply chain firms, PSE uses third-party credit ratings agencies to feed up-to-date financial data for all suppliers that appear on a programme's list of contracted organisations and tendering organisations. There are a number of firms in operation offering similar services, data sources and commercial indicators, including the financial strength of an organisation and their current or most recent history for meeting their payment obligations. These services allow the user to establish alerts that indicate when any of the suppliers being monitored on a programme breach the parameters that have been set.

Regularly updated financial information on the supply chain, together with the supplier utilisation model, allows for timely intervention to mitigate the risk of supplier insolvency. However, these indicators are of more significance for the critical sub-tiers of the supply chain than for the large tier 1 contractors.

Sub-tier supplier engineering

In order to select the most capable tier 1 contractors, a number of criteria are assessed as part of the initial pre-qualification tests, including financial capability and capacity, solvency, quality and previous performance on health, safety and environment issues. These tests and measures are then followed by a more focused and detailed examination of those suppliers that passed the initial PQQ stage, by further evaluating tenderers' responses against a project's specific requirements and their price for delivering the contract at the ITT stage.

This process can take many months to reach the point of contract award. The due diligence of ensuring that the most capable tenderer is awarded the contract is measured against the client's value-for-money criteria or most economically advantageous tender (MEAT). These values in turn form the framework of tests applied during procurement, with the MEAT and the strongest response, measured according to the balanced scorecard, being awarded the contract.

The need for due diligence in selecting a contractor is driven by the fact that the client is making a large capital investment. They therefore have a responsibility to test their suppliers in advance of awarding contracts. However, on large construction programmes the client very rarely directly contracts with the contractors who actually carry out the works. The tier 1 contractor directly contracted by a client procures tier 2 suppliers and subcontractors to deliver the work under their management.

Although the client may be making an investment using capital expenditure (CapEx) with a view to earning a long-term return, the tier 1 contractor views the same finance as part of their operational expenditure (OpEx). As a result, there is a difference in the manner in which the expenditure is managed by the client and by the main contractor. Operating expenditure is spent with due diligence, but that due diligence is not conducted to the same extent as required for capital expenditure. For example, during the preparation of a tier 1 tender, tier 2 suppliers may be engaged only to be asked to reduce their prices immediately following a contract award, with a view to enhancing the profit margin of the tier 1 contractor.

Moreover, the relationship between tier 1 contractors and their tier 2 subcontractors can change during a construction programme. For example, in the aftermath of the financial crisis of 2007/8 tier 2 suppliers were squeezed financially from two directions. The banks either withdrew credit or made it more expensive or difficult to secure, while the tier 1 contractors insisted on longer payment terms and lower prices from their supply chains. Given that tier 1 contractors' payment terms from their clients had not significantly changed, they were effectively using the client and supply chain as their own credit facility and at the same time insisting on discounts! Such actions tend to have a negative effect on the weaker organisations in the supply chain. All business organisations rely on the flow of cash and if that flow is restricted, companies can very quickly become insolvent.

While the total value of national and international sales of tier 1 contractors is very often equivalent to multi-million or even multi-billion pounds sterling, the turnover of many of their tier 2 suppliers range from only hundreds of thousands of pounds to multi-millions for the small to medium-sized enterprises. These smaller organisations are not large enough to employ a specialist finance director or dedicated finance team. As a result, the large corporations adopt stringent financial strategies to improve their cash flows, while the smaller suppliers continue to settle their invoices within the agreed terms, usually 28 days, and pay their workforce, usually weekly. Over time this disparity leads to cash-flow problems and financial stress, and in some cases insolvency and business failure.

These cash-flow issues are further exacerbated by the fragmented nature of the production process, which is spread through several tiers of different firms in the supply chain. Figure 9.4 highlights the paradox that very often the last in the chain of payment is the first in the chain to commit to expenditure, although the diagram in Figure 9.4 very much simplifies the flow of orders and payments. The right-hand arrows show the flow of orders and payments from the client down to the materials supplier. The arrows on the left indicate the flow of transactions for goods and services. For production to commence materials must be delivered by the tier 3 supplier, who must then wait for payment. The tier two subcontractor then assembles the materials on site and waits for payment from the tier 1 main contractor, who manages the project. The tier 1 contractor must then wait for payment by the client. Only after work has been signed off does the client pay the tier 1 contractor, who in turn pays the tier 2 contractor, who then pays the tier 3 supplier. As a result the tier 2 and 3 suppliers can often wait longer for payments than tier 1 contractors.

This model is a simplification of the system of pay-when-paid payment flows, assuming credit facilities are not involved to facilitate payment

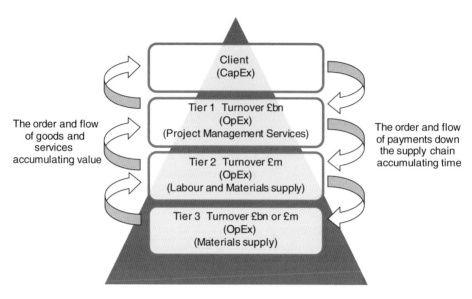

The order and flow
of goods and
services
accumulating value

**Client
(CapEx)**

**Tier 1   Turnover £bn
(OpEx)
(Project Management Services)**

**Tier 2   Turnover £m
(OpEx)
(Labour and Materials supply)**

**Tier 3   Turnover £bn or £m
(OpEx)
(Materials supply)**

The order and flow
of payments down
the supply chain
accumulating time

Figure 9.4   CapEx compared to OpEx – the paradox of expenditure.

in advance of receipt of payment. It is not a description of what occurs in most cases. However, this type of behaviour is driven by credit availability, which creates cash-flow difficulties and which in turn is a major cause of the financial failure of firms. For example, occasionally the following tier in the supply chain may require payment before the flow of payments reaches the tier above. This most often occurs at tier 2, when payment for goods is required before goods are delivered to site or on strict payment terms that can be enforced if further deliveries are needed. These payments may need to be made by the tier 2 contractor while still waiting for payment from the tier above. In these cases the tier 2 supplier relies on credit and other unrelated payments, which in a recession can be withheld or delayed to the detriment of the financial health of the tier 2 supplier.

The patterns of behaviour described in Figure 9.4 are usually conducted beyond the view or control of the clients of large construction programmes. However, PSE suggests that it should not be entirely beyond the scope of clients' control. PSE suggests that programmes require sensitive and targeted interventions to ensure the continuity of the programme, and the health of both the supply chain and the programme's micro-economic environment to foster success. Therefore the contracting strategy needs to focus on contract conditions, including payment provisions and how they flow down the supply chain. During tier 1 procurement the tier 1 contractors are questioned about their procurement and management policies, especially regarding their treatment of suppliers at tier 2.

According to PricewaterhouseCoopers LLP (2009), insolvencies in the UK across all industries in 2009 were higher than they had been in the previous 15 years. In construction, insolvencies continued at their peak rate of almost 700 per quarter. This poses a risk to programmes and projects, with the consequential effects of the insolvency of one firm on a programme being compounded in terms of cost and time. It is for these reasons that PSE is concerned with monitoring closely the sub-tiers of projects within a programme environment. Even a relatively small supplier can seriously affect the wider programme of which they form a part, if the firm is critical to delivery.

Case study

According to Lythaby and Mead (2012), 'Over £640 million of supplier risk was either removed or mitigated from the [London 2012] construction programme. Forty-three supplier insolvencies were avoided with zero impact, and while 11 were realised, their impact was minimised and managed through decisive early actions of the integrated ODA, delivery partner and main contractors.'

Identifying critical suppliers

To understand issues and risks lower down the supply chain, the programme supply chain team require detailed information from the tier 1 contractor, including their procurement schedule and budget values for each of their tier 2 package procurements. The programme supply chain manager can then work with the tier 1 contractor to agree which of the supply chain packages at tier 2 are critical to the delivery of the project. The initial list of critical packages is drawn up using a simple Pareto (80: 20) analysis of the packages. This is completed by ordering the packages from the highest to the lowest value, accumulating the values and drawing a line at the cumulative 80% mark. The top 80 per cent of packages by value are then automatically deemed critical to the project, owing to the financial contribution they make to the overall contract.

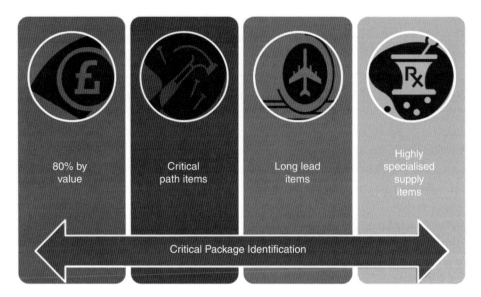

Figure 9.5 Identifying critical packages and their suppliers.

The tier 1 contractor, the client's project manager and the programme supply chain manager then review those packages in the remaining 20 per cent by value, to highlight any packages that fall into one or more of the following categories:

• packages that are on the project schedule's critical path for delivery, or
• packages with long-lead procurement items, or
• highly specialised packages from small supply markets, or
• items requiring an element of design.

The model for identifying critical packages is illustrated in Figure 9.5.

Having identified a list of critical packages, further information about each package is then needed, including the list of suppliers tendering for the packages (including their company registration numbers), and the budgeted cost flow for each package of work.

The information required should be readily available, as it should form part of the data the tier 1 contractor would need to assemble to assist in the process of delivering the contract. For example, the package procurement schedule and budget cost flows are important to the tier 1 contractors in establishing lead times and their own cash-flow forecasts. From experience of PSE implementation on large construction programmes, these items of data are forthcoming and made available on request, though not always without question or challenge. To reduce resistance to gaining access to this type of information, templates can be drawn up and included as works information items as part of the contract.

However, again from experience, some tier 1 contractors find it difficult to locate the company registration numbers of their tier 2 subcontractors, although almost all tier 1 contractors maintain that they run third-party financial credit checks against their proposed suppliers. To do so they must know the company registration number. They also need to know exactly which corporate entity they are entering into contract with, and so that information should be readily available. Nevertheless, it was invariably this item of information about critical suppliers that was the most difficult to obtain when implementing PSE. The difficulty could be indicative of a less-than-full due diligence financial assessment carried out by many tier 1 contractors during their own supplier selection processes.

Although only anecdotal, based on experience of working with many top UK and European tier 1 contractors, and where PSE has been implemented on a number of programmes, a number of issues regarding tier 2 contractors on tier 1 subcontractor lists have been highlighted, including companies that have long been out of business appearing on tender lists. There have also been examples of companies whose turnover is many times less than the contract value, and companies whose financial health is poor or, in some cases, very close to failure. These problems have come to light despite tier 1 contractors describing their own robust in-house processes for checking those aspects of their supply chain within their own tender submissions during their procurement by the programme procurement team. This suggests that the tests of due diligence are applied at the point of entry of a supplier on to a tier 1 preferred list or, at best, once on a preferred list, and revisited on an annual basis. These issues also suggest that much of this information may be held centrally at head office and that site-based managers do not have access to the relevant systems. It may be that those information systems are not utilised by the contractors' project-based teams, who may prefer to use their own site-based systems and in many cases choose to set up their own spreadsheets and forms specifically for the project in question, with little or no central intelligence gathering taking place.

For these reasons the exact same database as described earlier in this chapter for testing the capacity and financial health monitoring of tier 1 suppliers is applied to all critical sub-tier suppliers within the programme and across the multiple projects and packages. Once a programme begins on site, the PSE supplier database may contain many thousands of suppliers across multiple projects. This gives the PSE's supply chain function the ability to highlight risks associated with capacity, identifying where tier 2 suppliers might be tendering or supplying to multiple projects on the programme and could be in danger of overextending themselves. Interventions can then be made by the programme supply chain team to notify these potential pinch points to the tier 1 contractors' procurement teams, giving them a timely clarification and an opportunity to ascertain the level of risk and put appropriate remedies in place.

It is highly likely that tier 1 contractors know the exposure of each of their tier 2 suppliers to their own particular project. They may even possibly know the extent of the total exposure to the same tier 2 contractor across their business on other projects not related to the programme in question. However, what the tier 1 contractors cannot know is the exposure of any one of their tier 2 suppliers to the other tier 1 contractors on the programme. By having an overview of the whole programme, the PSE model can greatly assist in avoiding overheating certain suppliers or even sub-sectors, and can anticipate other possible difficulties such as cash-flow problems and the risk of contractor insolvency. This overview of the whole programme, and the positive and negative aspects of programme and project supply, are captured in Figure 9.6.

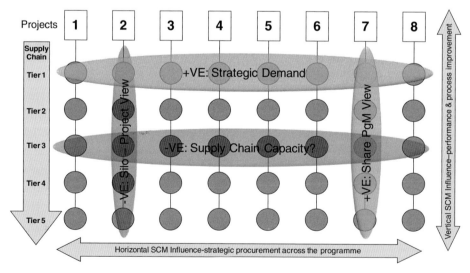

Figure 9.6   Programme supply chain management – dimensions of influence.

From the overview of a whole programme presented in Figure 9.6, PSE is seen as influencing supply for all programme projects both horizontally across projects and vertically in the supply chains. Eight projects within a programme are shown in Figure 9.6, each with supply chains consisting of five tiers. This forms a matrix that PSE uses to gain an overview for managing total demand strategically across all projects, while raising questions about the capacity of members of the supply chains to meet their commitments on different projects on which they may be working. Figure 9.6 also shows a project silo approach, which is seen as negative in Purchase and Supplier Engineering terms, but that can be counteracted with a shared management approach, which PSE both encourages and facilitates.

Concluding remarks

In effect the due diligence and tests that are applied during the selection and procurement of a supplier form the application of best-practice techniques applied with common sense and rigour. Management has the responsibility of satisfying itself fully that the terms and conditions of any agreement are practical and that the person carrying out due diligence is fully aware of all the essential factors involved in the transaction, from the capability and capacity of the seller to the financial robustness of the supplier to carry out the work.

It is inherent in the PSE model that the same level of due diligence is applied to the sub-tier critical elements of a project and its supply chain as it is to the direct tier 1 contract procurements. Critical issues need to be identified, as they are used to manage potential risks and avoid them if at all possible.

As it is not practical or feasible to assess the whole supply chain, a decision has to be taken about which elements to investigate. Certain issues may be determined by their relative size. For example, work packages are assessed by size, as are budgets and cash flow. However, some small work packages may be critical and have spill-over effects and consequences far greater than their size alone would imply. Using the Pareto principle, the great majority of problems tend to come from a relatively small number of relatively large sources, though judgement and knowledge may indicate the need to investigate smaller but equally critical packages.

Apart from size, other criteria for determining criticality may be based on the critical path, and in turn that may depend on the length of the lead time required by a firm or its material suppliers. Other factors that need to be taken into account are the level of technology used and the size of the supply pool. The level of technology might require a degree of specialism offered by only relatively few firms and that might reduce the supply pool, which is the number of firms capable of supplying the service. If the supply pool is restricted, contingency planning may be called for to identify alternative suppliers or strategies for dealing with problems in case they arise.

Finally, it is essential to take into account the risks arising across supply chains on a programme. Even if all tier 1 contractors apply the same due diligence tests during the procurement of their own suppliers, risks remain until the critical elements that impact upon delivery across a programme are noted and shared. Managing these risks is one of the functions of supplier engineering.

Capacity Management

Dos	Don'ts
Continually monitor the capacity of the critical supply chain (Tier 1 and below)	Assume the Tier 1s have the same market intelligence that you have
Agree and monitor the minimum level of financial risk that you/the client are prepared to accept from the supply chain	Intervene to the extent where you import risk
Monitor financial strength and look out for weakness, e.g insolvency risk, poor payment performance, large liabilities, CCJs, etc.	
Map the supply chain for each project across the programme	Assume the Tier 1s are doing this already
Identify and monitor the critical contracts, e.g. contracts of a high value, complexity or limited market that could impact on the successful delivery of the programme, including long-lead items and highly specialised services and products with small supply pools	
Maintain visibility within the supply chain	
Oversee critical supply chain package procurement at all appropriate tiers	

Capacity Management	
Dos	**Don'ts**
Intervene where appropriate/possible and enabled by your contracting strategy	
Influence where contractual intervention is not possible/practicable	
Upon identification of a weakness, make a positive intervention to avoid or mitigate the risk exposed	
Monitor lead times vs award	
Report risks as they materialise to appropriate owners	
Prepare capability reports by exception	
Review the flow-down of Balanced Scorecard and other contractual obligations into lower-tier supply subcontracts	

References

Lythaby, M., and Mead, J., (2012) 'Supply Chain Management – Insolvency Management', London 2012 Games construction project learning legacy paper, London, ODA.

OGC, (2008) *Supplier Financial Appraisal Guidance*, London, Office of Government Commerce.

PricewaterhouseCoopers LLP, (2009) *PwC Analysis: UK insolvencies at highest level for 15 years.* http://www.ukmediacentre.pwc.com/content/detail.aspx?releaseid=3083&newsareaid=2&zip=true (accessed 8 September 2012).

10

Performance management

The restored River Lea, once a dumping ground, is now an attraction for more natural visitors (photo courtesy of AECOM).

Programme Procurement in Construction: Learning from London 2012, First Edition.
John M. Mead and Stephen Gruneberg.
© 2013 John Wiley & Sons, Ltd. Published 2013 by John Wiley & Sons, Ltd.

The temporary seating 'wings' being built onto the Aquatics Centre. These added 2500 temporary extra seats during the Games (photo courtesy of Mark Lythaby).

View of the Olympic Stadium and Arcellor Mittal Orbit with their striking lighting during the Games (photo courtesy of AECOM).

Introduction

Earlier chapters have covered the need to motivate and retain the interest of firms in participating in a programme. They have identified the ability of firms to cope with the work in terms of their size and capacity. The previous chapter discussed the due diligence required to assess the financial strengths and weaknesses of the firms engaged to carry out work on different projects in a programme.

Earlier chapters have also shown that, based on the client's values and objectives, the PSE model begins by developing the client's requirements using a balanced scorecard. A framework for measuring performance can then be defined in terms of the criteria set out in the scorecard. This final chapter describes how those measures, which form the starting point for realising value, may be used to manage and improve performance during delivery. This completes the role of PSE in managing the delivery of the outcomes that were promised during the procurement of the projects across the whole of a programme.

The Purchase and Supplier Engineering model and programme management

The programme supply chain management function of PSE is designed to contribute programme-wide information and advice during delivery to reduce and, if possible, avoid risk. The actual contract delivery is left to the supply chain for overall management on a project-by-project basis. Nevertheless, performance management is an integral part of the contract delivery element of PSE, because the actual delivery by contractors can then be measured against their proposed performance set out by bidders at the tender stage.

Performance management is concerned with ensuring that the requirements of the client are identified, measured, reported and ultimately delivered in terms of the programme's packaging strategy, contracting strategy and supplier relationship management policies. Performance management also manages the collection and collation of data, information and knowledge from the supply chain to provide the basis of a legacy of 'lessons learned'. It might be hoped that if managed robustly, the knowledge gleaned from the procurement and delivery processes

Figure 10.1   Programme management model.

could be used to form the basis of improvement throughout all tiers of the supply chain.

Figure 10.1 illustrates a simplified programme management model, which is divided into three distinct elements. The first element is a document, the programme strategy, which sets out the requirements, governance, leadership, direction and structure of a programme. The second element is a team of managers, the programme management office (PMO), which defines the control, reporting and decision support processes to ensure the programme delivers the benefits expected from the stated requirements. The third element is a strategic approach to the management of programmes, Purchase and Supplier Engineering, (PSE), as described in this book. PSE is the element that oversees the process of delivering the appropriate goods and services as required to meet the client's and stakeholders' requirements.

These three distinct elements of programme management overlap to deliver assurance, where the PMO manages the programme within the governance and policies set by the programme leadership structure. The procurement strategy is set out through PSE as part of the programme strategy to manage the overall acquisition of goods and services through a packaging strategy, a contracting strategy and the programme supply chain approach. In Figure 10.1 performance management is at the interface of the PMO, which establishes a regime to manage the delivery of the projects by the contracted supply chains and PSE.

The PMO, PSE and the programme strategy need to be consistent with one another in order to realise the expected benefits of the programme. Establishing this alignment is often referred to as benefits realisation management.

Purchase and Supplier Engineering and the programme management office

This chapter shows how an aligned performance management regime can enhance the value already created by the PSE approach, in particular the effort expended at pre-contract stage, collecting information on suppliers and their own commitments to deliver the client's values and priorities written in their own tender submissions. Performance management reinforces benefits realisation management by combining the programme requirements and procurement objectives with the day-to-day management of the sub-programmes and projects.

One of the first tasks facing the PMO is to establish a programme management reporting and assurance structure as early as possible in the programme life cycle. This creates a governance structure focused on the client's requirements and the perceived benefits of delivering the programme. These requirements and benefits are developed in the early stages of PSE to become the 'DNA' of the programme, against which progress is measured and works procured.

Performance management within Purchase and Supplier Engineering

Performance management needs to be based on both quality management and continuous improvement: for example, the DMAIC cycle based on Define – Measure – Analyse – Improve – Control. The DMAIC process, illustrated in Figure 10.2, is a problem-solving approach facilitating management decision making based on data collected from the supply chain to eliminate variation in production and the quality of output.

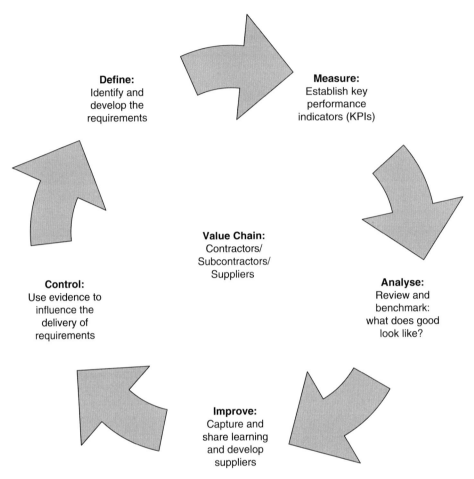

Figure 10.2 The cycle of Defining, Measuring, Analysing, Improving and Controlling for programme supply chains.

Analysis

The use of key performance indicators (KPIs) to monitor performance can highlight gaps in specific areas of delivery. These gaps in performance can then be further investigated to identify specific areas for improvement. The level of non-performance and the nature of the issues identified determine the type of corrective intervention required. This is often achieved through shared learning, with experience and knowledge drawn from the supply chain itself.

Action plans may be drawn up by the programme supply chain management team to make use of shared knowledge, while intervening, if necessary, with the help of required expertise for specifically identified problems, such as process improvement or value engineering. These action plans can be prepared, for example, using an RACI matrix. RACI is a systematic method of managing change effectively. The term 'RACI' is an acronym representing responsible, accountable, consulted and informed, which are the designations assigned to key individuals. The first step is to appoint an individual to be *responsible* and to take charge of the process. The second step is to identify a member of staff in authority, who must sign off any decisions and to whom the responsible person reports and is therefore *accountable*. Discussions need to involve others, who are *consulted* and participate in the process. Finally, those who may be affected by the change need to be *informed* of the decision. The implementation and subsequent performance change highlighted in the action plan can then be monitored by the programme supply chain manager responsible for the particular cluster or project.

In construction, the sharing of knowledge and expertise beyond the usual commercial barriers can help to improve delivery and operational processes. If members of the supply chain benefit from their experience of collaborative and integrated working, it can be argued that this could be seen as a legacy of the programme.

Control

Monthly operational data collected by the programme and project supply chain managers informs continuous gap analyses, by frequently comparing the difference between actual and expected or planned performance. The monthly monitoring at a project or sub-project level can then be complemented with quarterly meetings of projects or clusters to assess and compare performance. The process of measurement, monitoring and management in itself forms a virtuous cycle for continuous improvement and subsequent supplier development.

Monthly operational KPIs are also submitted to the programme supply chain managers by the contractors and their suppliers. The supply chain managers in turn report the headline findings on a regular basis to the client body to demonstrate the strategic delivery and progress of the programme. This data can also be published at a headline level to the programme's external communications team, possibly even on the programme's own website, to demonstrate transparency, manage expectations and motivate continuous improvement of all stakeholders. This process is part of the PMO's responsibilities and ensures consistency in all forms of reporting and communication.

Performance improvement through Purchase and Supplier Engineering

In many sectors of the economy, including construction, innovation takes place inside individual firms. These innovations are then seen and taken up by competitors in an adapted form. In construction, where the production processes are fragmented to a great extent, large programmes provide a particular opportunity for the industry to innovate. Major complex programmes, such as the BAA London Heathrow Terminal 5, London 2012 Games and the extremely complex Crossrail programme, can act as catalysts for change in construction, affecting both markets and suppliers.

One of the key outcomes of effective performance management is the early identification of trends and areas for improvement. It may be achieved with the help of clear, concise and relevant management information, which relates directly to the client's objectives for a programme or for their business. This applies particularly to the early identification of areas of the supply chain that need to be improved, and equally to areas of the supply chain that may be performing above expectation.

By measuring performance using simple, industry-accepted standards and indicators, such as the existing standard KPIs, the client can easily compare and benchmark performance against industry best practice. This in turn allows the implementation and dissemination of best practice throughout the supply chain.

To improve efficiency in organisations requires frequent measurement of various aspects of the production and organisational processes. As improvement implies change, there is a need to measure the current state of performance and compare it at some point in the future. A key part of the management of a culture of continuous improvement are the measures that show how the organisation and the process are progressing towards key productivity or quality gains.

However, measuring change alone is not sufficient to improve performance. Weighing a pig will not make it fatter! As the concept of Purchase and Supplier Engineering implies, a number of other factors need to be taken into account in the pursuit of innovation in construction. When agreeing the performance measures, the PMO needs to ensure that the measures are easily understood by all members of the supply chain and that they reflect the key areas for improvement. It is also important to align with the programme and client's objectives and ensure the measures are directly linked to agreed incentives and the commercial drivers of the programme. The measures need to make use of readily available 'business as usual' data, avoiding special or bespoke data, and where possible the measures should be representative of the programme as a whole, not just narrow financial or commercial aspects.

Measuring change is one of the functions of the performance management regime established jointly by the PSE and the PMO, who can identify appropriate sources of information, advise on the relevance of the data to achieving a desired end result, and assist in obtaining the appropriate level of knowledge required to make decisions. Another use of measuring performance is to encourage good practice through incentive schemes. Achievements can then be suitably rewarded, with a view to motivating individuals to make further improvements at an individual and organisational level. To be credible, the measures should also openly reflect and highlight when performance is not adequate or acceptable.

Benchmarking

Benchmarking can provide a direct comparison for similar (and competing) businesses to examine their own performance in a number of given areas; it can identify opportunities for improvement within a firm and often provides a catalyst for implementing change. For example, this can be achieved by utilising the KPIs, established as part of the performance measurement regime beyond the useful financial criteria. The collection and analysis of data on key areas, such as sustainability (and in particular the 'triple bottom line' approach) and standard health and safety metrics (such as accident-free days) are also used to measure and benchmark performance across the UK construction industry.

Concluding remarks

A set of performance measures and a Performance Management system for delivering improvement need to be established early in a programme, as they underpin how the client's vision is realised. This is achieved using scorecards to measure particular aspects of performance. Scorecard measures need to align to the programme vision and these measures become a core part of the procurement strategy. Once a performance management regime has been established, the data collected through the supply chain can help to identify trends and weaknesses in the market with individual suppliers and even in the supply chains themselves.

Performance management and gap analysis information can be used to improve the performance of individual suppliers and supply chains, but it needs to be focused and specific, using reliable and widely understood information such as construction industry or standard key performance indicators.

Performance management can be used not only to monitor progress, but to enable benchmarking, improve performance, encourage innovation and reward achievement as part of incentive schemes. Benchmarking can be an effective method of helping businesses compare performance and use 'lessons learned' to instigate improvement initiatives. It is entirely appropriate to conclude this account of PSE on a note of optimism concerning not only its ability to make improvements on the programmes, where it is used as an approach to management, but also its ability to make a potential contribution to change and innovation in the construction industry in general.

Contract Delivery	
Dos	**Don'ts**
Performance Management	
Ensure project team and suppliers understand the performance management mechanism, system and reporting requirements	
Use data that emerges from standard processes and is part of natural delivery mechanisms	Create bespoke data requirements
Contract administration	
Implement contract administration code to ensure a consistent approach across contracts	
Build contract admin team early and involve in procurement process to ensure familiarity and minimise handover requirement	
Promote resource continuity by key procurement management staff moving on to perform post-contract roles	
Provide an appropriate and bespoke level of training for all members of project team responsible for post-contract delivery	

Index

accessibility 129
acid test, the 180
action plans 198
appetite of firms 21, 105, 117, 121, 123, 159, 160
Aquatics Centre 83, 124
architects 38
assurance review 141, 144
auditability 129

BAA London Heathrow Terminal 5 200
balanced scorecard 13, 47, 53, 55, 63, 98, 107, 110, 129, 158, 182, 195
basketball 84
benchmarking 101, 201
benefits
   matrix 98
   realisation management 197
BMX 84
board reports 146
bounded rationality 22
bulk purchasing 122
business case 96, 98
business opportunities 115–116
BusinessLink 118
buyers' market 23, 173
buying club 101

call out flags 166
capability 123
capacity
   sector 94
   suppliers' 71, 117, 123, 172
capital investment 182
cash flow 183
CH2M Hill 10

challenge, risk of 132, 142, 159
Chambers of Commerce 118
CIOB 37, 40
Civil Engineering Contractors' Association 125
civil engineering
   contracts 162
   works 84
client
   body 31
   requirements of 8
   role of 22, 24, 30
   sponsor representatives 101
   supply framework 101
   team 25, 26
CLM 10
clustering 67
   packages 77
clusters 78, 162
Code of Practice for Project Management 36
code of practice, procurement 141
common component
   procurement strategy 100
   strategy 96
common components 90
   and commodities 162
common performance specification 99
company registration numbers 186
compensation event process 80
Competefor 78, 118
competition 123
Competition Act, 1998 26
competitive dialogue 84
competitive tension 105, 160
conflict 35
conflicting projects 26

Programme Procurement in Construction: Learning from London 2012, First Edition.
John M. Mead and Stephen Gruneberg.
© 2013 John Wiley & Sons, Ltd. Published 2013 by John Wiley & Sons, Ltd.

Keep up with critical fields

Would you like to receive up-to-date information on our books, journals and databases in the areas that interest you, direct to your mailbox?

Join the **Wiley e-mail service** - a convenient way to receive updates and exclusive discount offers on products from us.

Simply visit **www.wiley.com/email** and register online

We won't bombard you with emails and we'll only email you with information that's relevant to you. We will ALWAYS respect your e-mail privacy and NEVER sell, rent, or exchange your e-mail address to any outside company. Full details on our privacy policy can be found online.

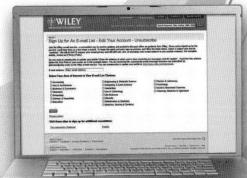

17841